Frank Clowes, Alexander Cruikshank Houston

Bacterial Treatment of Crude Sewage

Second report

Frank Clowes, Alexander Cruikshank Houston

Bacterial Treatment of Crude Sewage
Second report

ISBN/EAN: 9783337315122

Printed in Europe, USA, Canada, Australia, Japan

Cover: Foto ©berggeist007 / pixelio.de

More available books at **www.hansebooks.com**

London County Council.

BACTERIAL TREATMENT OF CRUDE SEWAGE

(SECOND REPORT).

BY

Dr. CLOWES AND Dr. HOUSTON.

EXPERIMENTAL INTERMITTENT TREATMENT

OF

LONDON CRUDE SEWAGE

IN

THE COKE-BEDS AT CROSSNESS.

PRESENTED BY

PROFESSOR FRANK CLOWES, D.Sc. (Lond.), F.I.C.
(Chief Chemist to the Council),

To the Main Drainage Committee of the Council.

JAS. TRUSCOTT AND SON, PRINTERS, SUFFOLK LANE, CANNON STREET, E.C.

DIVISION I.—CHEMICAL AND GENERAL.

BY

PROFESSOR FRANK CLOWES, D.Sc. (LOND.), F.I.C.

(Chief Chemist to the Council).

CONTENTS.

DIVISION I.—CHEMICAL AND GENERAL.

SECTION I.—GENERAL OBJECTS AND CONCLUSIONS.

1. Introductory.
2. Objects of the Coke-bed Experiments.
3. General results obtained.
4. General Conclusions.
5. Advantages of Bacterial over Chemical Treatment.

SECTION II.—CONSTRUCTION AND WORKING OF THE COKE-BEDS.

1. Construction and Details of the Coke-beds.
2. Method of Working the Beds, and their present condition.
3. History of each Coke-bed.
4. Experimental proof of the Aëration of the Coke-beds.
5. Variation in the Condition of the Raw Sewage, and its effect upon the Effluent.
6. Comparative purity of clear Sewage, Bacterial Effluent and Chemical Effluent.

SECTION III.—TABLES AND CURVES, SHOWING THE RESULTS OF THE CHEMICAL EXAMINATION OF THE CRUDE SEWAGE BEFORE AND AFTER TREATMENT IN THE COKE-BEDS.

Table I.—Results obtained by the Chemical Determination of the Relative Amounts of Dissolved Oxidisable Matter in the Raw Sewage before and after the Bacterial Treatment.

Diagram 1.—Curves showing the Variation in the Relative Amount of Dissolved Oxidisable Matter in the Crude and in the Treated Sewage from July, 1898, to February, 1899, and the Rainfall at corresponding times. Also Curves showing the Variation in the Weekly Averages of Percentage Purification effected on the Raw Sewage by the Bacterial Treatment, as measured by the removal of Dissolved Oxidisable Matter.

Diagram 2.—Curve showing the Daily Variation in the Percentage Purification effected on the Raw Sewage by the Bacterial Treatment.

(The Diagrams will be found immediately following page 41 of the text.)

DIVISION I.—CHEMICAL.

SECTION I.

GENERAL OBJECTS AND CONCLUSIONS.

1.—INTRODUCTORY.

After duly considering the different methods of sewage treatment which have been at present proposed (see pages 19, 20), it appeared to me that the intermittent treatment in bacteria coke-beds was the one most likely to be suitable to the conditions at the Council's Northern and Southern Outfalls. Much information has already been laid before the Main Drainage Committee on this method, but chiefly with respect to the results obtained by treating the *effluent* from the chemical precipitation of London sewage. I considered that further information and experience was desirable before the Committee could be advised to embark on the scheme of treating the whole or even a considerable portion of the raw sewage by this process, which is usually called that of "intermittent filtration through coke-beds." This treatment undoubtedly depends upon the action of bacteria, and the term "filtration," which implies mechanical removal of suspended impurity should be abandoned.

On the 15th of February, 1898, the Council approved, on the recommendation of the Main Drainage Committee, my proposal that further experiments should be made on this process of intermittent bacterial treatment in coke-beds, and the chief engineer, Sir Alexander Binnie, willingly accorded his cooperation. Dr. Houston was retained for the purpose of conducting bacteriological investigations during the first year of the experimental work, and a grant was voted by the Council for the purpose of laying down coke-beds on a small scale at Crossness and for maintaining their working. Thanks to the cooperation of Mr. E. J. Beal, the superintendent, and of Mr. J. W. H. Biggs, the assistant chemist at Crossness, these experimental coke-beds have been constructed and worked to my entire satisfaction.

In the present report the details of construction and the method of working the coke-beds are explained, and the general results arrived at, after about ten months experience of the filtration, are stated.

It appeared to me to be necessary that the crude sewage at the Northern Outfall should be subjected to similar experimental treatment, and a further sum was accordingly voted by the Main Drainage Committee on 21st July, 1898, to equip small experimental coke-beds at the Northern Outfall. The full particulars of these coke-beds and of the results obtained by them, will be included in a special report in due course.

The Committee has already received, in a Report which they ordered to be printed and circulated on the 16th June, 1898, the results obtained by Dr. Houston from a bacteriological examination of the crude sewage as it arrives at Crossness and at Barking. In the present Report the results are stated, which were obtained by Dr. Houston from a bacteriological examination of the Crossness sewage after its passage through the coke-beds (pages 17 to 41).

It should be explained that the experiments on intermittent treatment in the coke-beds have been carried out by filling the coke-beds with sewage until the coke is just submerged, then allowing the sewage to remain in contact with the coke for several hours, and finally draining the liquid away completely. The coke is then allowed to remain for some time in contact with the air, which fills the interstices of the coke-bed when the liquid has flowed away. This series of processes is repeated at regular intervals.

Fuller details as to the construction and working of the coke-beds will be found in Section II. (pages 7 to 10) of this Report.

2.—OBJECTS OF THE SO-CALLED "FILTRATION" EXPERIMENTS.

The process of "filtration" of raw sewage, or its bacterial treatment in coke-beds, is already in use for purifying the sewage of several towns containing a population less numerous than that of London. But in order to render this process adaptable to the treatment of London sewage it appeared to be necessary to use coke-beds which were deeper and which could be more rapidly filled and emptied than those usually employed elsewhere. And further, since London sewage is of a special character, consisting not only of domestic drainage but of manufacturing refuse of the most varied description, it seemed advisable to ascertain by a lengthened experience whether its purification was possible, and whether it could be regularly maintained by the process of bacterial treatment. With these objects in view, coke of fairly large size was employed, and the coke was introduced into tanks of considerable depth so as to enable the layer of coke to be progressively increased in thickness during the progress of the experiments.

The main objects of the experiments were to ascertain—
    (a) The effect of using the coke in fragments about the size of a walnut.
    (b) The effect of increasing the depth of the layer of coke beyond the usual limit.
    (c) The extent to which the raw sewage underwent purification by the treatment.
    (d) The practicability of maintaining the constant passage of raw sewage through the same coke-bed, without deterioration occurring, either in the bed or in the effluent.

(e) The amount of sewage which could be treated daily by a superficial unit of the coke-bed.
(f) The extent to which the effluent underwent further improvement by its passage through a second similar coke-bed.
(g) The suitability of the effluent for maintaining the life of fish.
(h) The effect of the treatment on the number and nature of the bacteria which were present in the raw sewage.

3.—GENERAL RESULTS OBTAINED.

It may be stated that the general results obtained from about 10 months' experience of the coke-bed treatment point to the following conclusions—

(a) Size of coke.

The use of ordinary gas coke, in pieces about the size of walnuts, seems to be attended with the following advantages, as compared with the use of smaller coke. The larger coke enables the bed to hold a larger volume of sewage. The beds now in use had an original capacity for sewage which was nearly equal to the volume of the coke which they contained, in place of only 20 or 30 per cent. of that volume, as is shown by beds containing smaller coke. The use of the larger coke also allows the bed to be more rapidly filled and emptied, and to be more completely emptied and aërated.

(b) Depth of coke.

Coke-beds similar in character, but differing in depth, have been found to give practically identical purifying effects. At present experience has been obtained only with a four-foot and with a six-foot bed; but I am arranging to extend these depths considerably.
[June 21st, 1899—A bed 13 feet in depth has now been working satisfactorily for over nine weeks, and has given a purification approximately equal to that effected by the four-foot bed.]

(c) Chemical purification effected by a single treatment.

The sewage had been roughly screened before reaching the coke-beds, and it was free from larger matter usually described as "filth," and from coarse sand and heavy mineral road detritus, but it contained all the suspended solid matter usually termed "sludge."

During part of the period reported on, the coke-beds received only one charge of this raw sewage daily; but latterly all the coke-beds have received two charges daily of raw Crossness sewage.

The coke-beds have removed the whole of the suspended matter or "sludge" from the crude sewage; and they have yielded an effluent which occasionally shows a slight turbidity, apparently due in ordinary flow, mainly to the presence of bacteria, but which is increased in storm flow by fine clay and mud.

Not only has the suspended matter been removed, but the removal of the dissolved oxidisable and putrescible matters of the raw sewage has been secured to the average extent of 51·3 per cent. by the single process, the four-foot coke-bed giving 52·7 and the primary six-foot coke-bed 49·9 per cent. The effluent thus produced remains free from objectionable odour when it is kept in open or in closed vessels, provided the bacteria present in it are not removed or killed by special subsequent treatment. This effluent could, therefore, produce no offensive smell when it is introduced into the river.

(d) Permanency of coke-beds.

Since the coke-beds have become regular in their action, neither the effluent nor the coke itself has become foul. The coke which has been in use for upwards of 12 months has shown no general tendency to break up, but the surface of each piece of coke has become partially covered with soft matter which consists mainly of fine particles of coke with some fine sand, woody and vegetable tissue, cotton and woollen fibres and diatoms. Below this thin film the coke is perfectly hard and no sand has penetrated two millimetres below the surface. The sand when examined by polarised light was found not to be granite sand derived from road detritus. The ash in the coke has been reduced in amount by about 25 per cent. during its exposure to sewage in the coke-bed.

The capacity of the four-foot coke-bed, has during the period in review been reduced from 50 to 33 per cent. of the whole volume of the bed, and this reduction of capacity appears to be mainly due to fragments of straw and chaff, apparently derived from horse-dung, and to woody-fibre derived from the wear of wood pavements. The treatment of raw sewage by this coke-bed is being continued in order to ascertain whether the decrease in the capacity continues. It has been ascertained that the original capacity is not restored in any degree by prolonged aëration, which proves that the deposit on the coke surface was not organic matter or animal origin; but it has been found that the vegetable tissue, which appears to be the main cause of the decrease in capacity, can be in great measure separated from the raw sewage by a brief period of sedimentation before the sewage is allowed to flow into the coke-bed.

(e) Amount of sewage which can be treated daily by a superficial unit of the coke-bed.

The volume of sewage which can be passed through the coke-bed per unit of superficial area has not yet attained its maximum, since the depth of the coke-bed is being further increased. It originally amounted to 555,000 gallons per acre per day for the four-foot coke-bed, and to 832,500 gallons per acre per day for the six-foot coke-bed. This represents in each case one filling per day; but, as has been already stated, two fillings have been made successfully for six months, and this corresponds to 1,665,000 gallons per acre per day for the six-foot coke-bed. These amounts are reduced after ten months' working of the coke-beds to 370,000 gallons per acre for a single filling of the four-foot coke-bed, and to 673,400 for the six-foot coke-bed.

The maximum possible rate of treatment by each coke-bed is possibly not yet reached. This is an important point to be determined, since the practicability of applying this method to the treatment of the whole of the London sewage depends largely upon the superficial area which will be required for the laying down of the coke-beds. The above daily rate of treatment will naturally be augmented as the depth of the coke-bed is increased, and if the satisfactory working of the 13-foot bed is maintained, it will treat a volume of raw sewage equal to at least three-and-a-half million gallons per day.

(*f*) *Secondary treatment.*

It has been stated that the purification effected by a single treatment of the raw sewage in the coke-beds amounts to a complete removal of the suspended matters, and to a further removal of an average of at least 51·3 per cent. of the dissolved putrescible oxidisable matter. The primary six-foot coke-bed actually removed on the average 49·9 per cent. of dissolved impurity, and a second process has effected thus far an additional purification of about 19·3 per cent., giving a total average purification of the clarified raw sewage amounting to about 69·2 per cent.

(*g*) *Effect of the effluent on fish.*

Fish die at once when they are placed in the present effluent produced by chemical precipitation, probably because there is a serious deficiency of dissolved oxygen in the impure liquid, and therefore their respiration cannot be maintained. Not only goldfish, but roach, dace and perch have lived for months in the first effluent from the coke-beds, and they apparently would live and thrive in this liquid for an indefinite period.

(*h*) *Bacteriological character of the effluent.*

The results obtained by the bacteriological examination of the effluent by Dr. Houston (see pp. 17 to 41) thus far seem to indicate that the coke treatment does not by any means remove the bacteria from the crude sewage, and indeed does not materially reduce the number. It shows that the presence of many of the bacteria in the effluent is possibly unobjectionable, and is probably necessary for the purpose of completing the purification of the effluent when it has flowed into the river; but it further shows that some of the bacteria whose presence might be looked upon as undesirable in drinking water pass through the coke-beds.

4.—GENERAL CONCLUSIONS.

The above considerations show that neither on chemical, not possibly on bacteriological grounds can any serious objection be raised to the introduction of the effluent from the coke-beds into a portion of the river Thames which is cut off by locks from the Intakes of the Water Companies, and the water from which is not employed for drinking purposes, and cannot be used for drinking on account of its "brackish" nature. The effluent certainly will not cause any deposit upon the river-bed, and will even tend to render the turbid water of the lower river more clear and transparent. At the same time, the liquid discharged from the outfall into the river will be sweet and entirely free from smell. Further, it will carry into the river the bacteria necessary for completing its own purification in contact with the aërated river water, and under no conditions can it therefore become foul after it has mingled with the stream. The effluent will in no way interfere with fish-life in the stream.

5.—ADVANTAGES OF BACTERIAL OVER CHEMICAL TREATMENT.

As compared with the present process of chemical precipitation and sedimentation, the bacterial process presents the following advantages—

(*a*) It requires no chemicals.
(*b*) It produces no offensive sludge, but only a deposit of sand or vegetable tissue which is free from odour.
(*c*) It removes the whole of the suspended matter, instead of only about 80·0 per cent. thereof.
(*d*) It effects the removal of 51·3 per cent. of the dissolved oxidisable and putrescible matter, as compared with the removal of 17 per cent. only, effected by the present chemical treatment.
(*e*) Further, the resultant liquid is entirely free from objectionable smell, and does not become foul when it is kept; it further maintains the life of fish.

SECTION II.

CONSTRUCTION AND WORKING OF THE COKE-BEDS.

1.—CONSTRUCTION AND DETAILS OF THE COKE-BEDS.

For the purposes of the present experiments three brick-lined tanks with brickwork bottoms were specially provided with suitable outlets and with means of charging them with raw sewage by means of a pump and distributing troughs. A series of parallel loose-jointed stoneware drain pipes were laid on the bottom of the tanks, and were connected with suitable outlets; this arrangement served to draw off the effluent.

Two of these tanks (A, B) are precisely similar in dimensions, being 22 feet 6 inches long, 10 feet 8 inches wide, and 12 feet deep, and the superficial area of each is $\frac{1}{73}$rd of an acre. The third tank (C) is of less regular shape, but of the same area as (A) and (B), and 6 feet in depth.

The Single Coke-bed.

One of the tanks (A) was filled to a depth of four feet with coke of uniform size, each fragment being about as large as a walnut. The coke was drawn from the retorts of the gasworks at the Crossness Outfall, and was found by long soaking in water to absorb 15 per cent. of its weight of water. The coke-bed had a sewage capacity of 3,000 gallons, which is equal to 50 per cent. of the volume occupied by the coke and air space; the volume of the coke amounts to 960 cubic feet.

This coke-bed is best described as "The Single Coke-bed," since it has been used for the purpose of subjecting the sewage to a single process only.

The Double Coke-bed.

The other two tanks (C, B) have been filled to a depth of six feet with precisely similar coke, which was obtained from the South Metropolitan Gasworks. The sewage capacity of each of these tanks is 4,500 gallons. As compared with the single coke-bed, the capacity of the bed is therefore increased in proportion to the increased volume of coke which it contains.

Tank C is at a higher level than tank B, and the two coke-beds have been, therefore, easily rendered available for a process of double treatment. The raw sewage has been pumped into tank C, and the effluent from this tank has been allowed to flow by gravitation into tank B and there to undergo a second treatment.

In order to avoid confusion the two tanks, C and B, have been called "The Double Coke-bed," and for purposes of distinction tank C has been called the "Primary," and tank B the "Secondary Coke-Bed."

It should be stated that the effluent from the Primary Coke-bed was passed through a small laboratory coke-vessel pending the preparation of the Secondary Coke-bed. This laboratory coke-vessel consisted of a glass bottle of 2·75 gallons capacity, which was filled with small coke, and which had a sewage capacity of 1·25 gallons, or about 45 per cent. of that of the coke which it contained.

2.—METHOD OF WORKING AND PRESENT CONDITION OF THE COKE-BEDS.

After each coke-bed has been filled from above to the level of the upper surface of the coke, a process which occupies about seven minutes for the four-foot bed, the sewage remains in contact with the coke for three hours. The liquid is then allowed to slowly flow by gravitation from the bottom of the coke-bed, the process of emptying requiring an hour for the four-foot bed, and the coke-bed is then allowed to stand empty for about eight hours in order that the surface of the coke fragments may become aërated. In the case of the secondary coke-bed the aëration process lasted only seven hours.

The coke-bed requires to be daily filled and emptied and aërated for about four weeks before it is "matured" and begins to exert its full purifying effect upon the sewage. During this interval the coke is doubtless becoming sown with bacteria, which are the active purifying agents, and which are present in large numbers in the raw sewage. This process of "maturing" the coke-bed has usually been carried on during its construction, by constantly treating the coke with sewage while it is being introduced into the tank.

In the earlier experiments with the "single coke-bed," when it was being filled twice a day, and before it was fully "matured," the coke-bed evidently became overworked, and accordingly returned a foul effluent, and the coke itself became foul. A fortnight's rest in an empty condition restored its activity, and it has not again given any unsatisfactory results whilst being filled with sewage once a day, and latterly when it has again been filled twice a day.

The coke-beds are not filled on Sundays. They have also rested for four consecutive days, including Sunday, during the Whitsun recess, and on July 30th and August 1st, 5th, 6th and 7th, the latter rest being partly due to the necessity of repairing the sewage pump. The coke-beds rested on five other days. With these exceptions the coke-beds have been continuously at work since the date when they were started.

The surface of each working coke-bed is broken up to a depth of several inches twice a week by being raked over; this keeps the surface open, and there is no appearance of its becoming clogged.

At longer intervals a hole extending from the surface of the coke-bed to the bottom of the tank has been dug in the coke. The whole of the coke has on every occasion been found to be perfectly sweet when it has been thus exposed, and possesses only a slight earthy odour. The surface of the upper portion, to a depth of three or four inches, is not quite bright, but it emits no foul smell.

The use of comparatively large pieces of coke in constructing the coke-bed enables the bed to hold a charge of sewage which is considerably greater than that which would fill a similar bed constructed of the smaller coke which was formerly employed. A single daily filling of the coke-bed deals, therefore, with almost as large an amount of sewage as the double daily filling of an ordinary bed. The larger size of the coke will doubtless also admit of the coke-bed being considerably deepened without losing the possibility of fully aërating the coke. It may be confidently anticipated that the larger coke and greater depth of the bed will thus enable the necessarily large amount of sewage to be dealt with.

Since the single coke-bed settled down to a steady rate of working, the number of daily fillings has been increased to two, and the depth of one of the coke-beds has been increased from six to thirteen feet in order to ascertain the possibility of working with deeper coke-beds.

3.—HISTORY OF EACH COKE-BED.
The Single Coke-bed.

The single coke-bed was filled with coke to a depth of of 4 ft. It was first charged with sewage on April 22nd, and from that date until June 23rd it was charged with crude sewage twice daily, with the exception of Sundays when it rested entirely, and of Saturdays when it received one filling only.

On June 23rd it was decided to give the coke-bed complete rest for a fortnight, since both the bed and the effluent were becoming foul. There is little doubt that this foulness arose from the fact that undue work was thrown upon the bed before it had become "matured." The impurities, therefore, gradually collected in the coke, since they could not be dealt with by the bacteria, which were not established in the bed at that period in sufficient number. Ultimately the impurities accumulated in the bed in such quantity as to render the purification insufficient. There is no doubt that this state is unlikely to arise in a coke-bed which is not overworked in its immature state.

After the fortnight's rest, the bed was filled only once a day, and effected satisfactory purification of the crude sewage continuously until November 7th, 1898.

During this period the coke-bed was stopped on certain public holidays and for the repair of the sewage pump, these stoppages amounting in all to 12 days.

On November 8th, 1898, a commencement was made with two fillings a day, and this was maintained until February 18th, 1899. The results obtained by two fillings have been perfectly satisfactory.

Up to the present date (February 18th) this coke-bed has been charged with raw sewage 339 times, and the one layer of coke has dealt with about 847,500 gallons of the raw sewage. The coke-bed has, therefore, removed from the sewage an amount of solid matter which in the dry condition would weigh 32·4 cwt. This solid matter would represent 20·25 tons of sludge, containing 92 per cent. of moisture; or taking one ton of sludge as occupying 33 cubic feet, the sludge removed by the coke-bed would fill the empty coke-bed tank to a depth of 2 feet 9 inches.

The Primary Coke-bed for the first stage of Double Treatment.

This coke-bed was charged with coke somewhat slowly owing to the difficulty of procuring coke at the time when the bed was being constructed. From July 12th to September 1st, the period over which the introduction of coke extended, the coke-bed was, however, frequently charged with crude sewage, in order to carry on the "maturing" process during the lengthened process of construction. The bed started its regular work in the treatment of crude sewage on September 1st, and from that date until February 18th it has been filled 213 times.

This coke-bed, which is six feet in depth, is now producing an effluent in which the purification is practically the same as that effected by the single coke-bed, which is four feet in depth. It may therefore be inferred that the purification effected by the bed will not be diminished when its depth is still further augmented.

The Secondary Coke-bed for the second stage of Double Treatment.

The Secondary Coke-bed was "matured" by treatment with sewage during its construction. It received its first charge of raw sewage on June 21st, and was worked until August 31st as a single crude sewage coke-bed. During that period it received 60 fillings of raw sewage, and 244,200 gallons of sewage in all were passed through it. It removed during this period an amount of solid matter from the sewage, which in the dry condition would weigh 9·35 cwt. This solid matter would be equivalent to 5·85 tons of sludge containing 92 per cent. of moisture.

Since September 1st this coke-bed has been used as a secondary bed, and has received the effluent from the primary coke-bed and subjected it to a second process of treatment. Since the primary effluent is usually clear, this secondary bed has been spared the process of removal of solid suspended matter and has exerted all its purifying action upon the dissolved oxidisable matter. Since the secondary coke-bed has been in operation the use of the small laboratory vessel has been stopped.

4.—Experimental proof of the aëration of the coke-beds.

In order to ascertain whether the surface of the fragments of coke became fully aërated throughout the bed during the successive chargings with raw sewage, two vertical pipes were inserted into the bed reaching to the depths of 6 feet and 13 feet respectively. After the sewage had flowed away from the bed samples of air were drawn off from the interspaces between the coke fragments at stated intervals, and the percentage proportions of oxygen and carbonic acid were estimated in this air. The results, which in the case of the 13-foot depth were only of a preliminary character, indicated that even after the air had been in contact with the lower strata of the coke for seventy hours, the air still contained an average of about 75 per cent. of its original oxygen, and the average amount of carbonic acid did not exceed 3 per cent., as is indicated by the tabulated results below.

This evidently represents an entirely satisfactory condition of aëration of the coke surfaces.

Six-foot depth.			Thirteen-foot depth.		
Number of hours since sewage drained off.	Percentage of oxygen in air.	Percentage of carbonic acid in air.	Number of hours since sewage drained off.	Percentage of oxygen in air.	Percentage of carbonic acid in air.
4	19·8	0·4	22	18·4	1·4
22	9·8	5·8	26·75	14·0	3·8
24·5	10·0	6·0	50·75	14·8	3·0
37	17·8	2·0	51·25	15·3	3·3
40·5	16·8	2·4	70	14·7	0·8

5.—VARIATION IN THE CONDITION OF THE CRUDE SEWAGE AND ITS EFFECT UPON THE EFFLUENT.

The sewage which has been subjected to the above treatment has varied much in character as judged by the amount of suspended matter and of dissolved oxidisable matter which it contains. The effluent derived from this sewage appears as a rule to have undergone a variation corresponding to that of the sewage from which it has been produced. The results of the daily estimations of oxidisable matter in the crude sewage and in the corresponding effluent will be found in the Diagrams and Tables in Section III.

In order to render the variations in the degree of purification effected by the coke-beds more evident, the percentage daily purification has been drawn out as a curve in Diagram I., and in the same diagram the average weekly percentage purification has been represented and the relative amounts of oxidisable matter in the crude sewage and in the effluent also have been treated as curves. It will be understood that the results deal only with the dissolved oxidisable matter in the sewage and in the effluent. The solid suspended matters of the sewage wholly disappear in the coke-bed and are therefore not taken into account.

The occasional diminution in the percentage purification does not appear to be due to an acid reaction of the sewage hindering the bacterial action, since the sewage is always either alkaline or neutral in reaction. Neither is this diminished purification apparently to be referred to the presence in the sewage of undue proportions of chemical refuse derived from gas works and chemical works. No evidence has been obtained of interference with the normal action of the coke-bed from such causes.

It appears, however, very probable that the purification of the raw sewage is most complete when the sewage is in a fairly dilute and fresh condition. It is of interest to consider the rainfall curve on the diagram in this connection. The least satisfactory results were obtained during the hot dry weather of the recent summer, when the sewage arrived at Crossness in a less dilute condition, owing to the absence of rain, and in an offensive and putrescent condition owing to the high temperature of the air. Even when the sewage was poured upon the coke-beds in this offensive condition, the effluent from the beds was not offensive in character, and it did not become offensive when it was kept. It differed from better effluents only by containing a larger amount of dissolved oxidisable matter.

6.—RELATIVE PURITY OF CLEAR SEWAGE, CHEMICAL EFFLUENT, COKE-BED EFFLUENTS, AND LOWER RIVER WATER.

The relative amounts of dissolved putrescible matter in the sewage, the chemical effluent, and the coke-bed effluent, as measured by the oxygen which they absorb from permanganate, are as follows:—

	Impurity of liquid.	Percentage purification calculated on raw sewage.
Raw sewage	3·696	—
Chemical effluent	3·070	16·9
Coke-bed effluent (single treatment)	1·799	51·3
Coke-bed effluent (double treatment)	1·137	69·2
River water, high tide	0·350	—
River water, low tide	0·429	—

A comparison of these numbers with one another shows that by substituting a single coke treatment for chemical treatment, the effluent sewage discharged into the river would be completely free from suspended impurity, and would possess a purity, as regards dissolved putrescible matter, of 51·3 as compared with 16·9 in the present effluent, representing an improvement of 67·1 per cent. If discharged after double treatment in the coke-beds the percentage improvement on the chemical effluent would be 75·6. The bacterial action continuing in the river would rapidly bring the purity of such a liquid into a condition equalling that of the river water itself.

SECTION III.

TABULATION OF RESULTS OF CHEMICAL EXAMINATION OF THE RAW AND THE BACTERIALLY-TREATED SEWAGE.

The results of the examination of the sewage before and after its passage through coke-beds were obtained by estimating the amount of dissolved oxidisable matter in the crude sewage, which had been clarified by passage through a paper filter, and then making a similar estimation in the same sewage after it had passed through the coke-bed.

The oxidisable matter was estimated by exposing the liquid, which had been clarified by filtration if necessary, to the action of acidified potassium permanganate, in closely stoppered bottles, for four hours at a temperature of 80° Fahr.

The oxidisable matters are measured by stating the number of grains of oxygen per gallon of the liquid which are required for their oxidation. When the coke-bed has been filled twice in a day, a sample of each effluent is collected, and these equal samples are mixed for the chemical estimation.

This method of estimation is not appreciably affected by the small amount of nitrite which is occasionally present in the effluent.

The Table which includes the numerical results will be found on pages 11 to 14. The results obtained will be more readily appreciated by a consideration of the curves in the Diagrams which immediately follows page 41.

TABLE I.—SINGLE COKE-BED (4 FEET).

Date	Number of times filled daily.	Number of grains of oxygen absorbed in 4 hours by one gallon of	
		Raw sewage.	Effluent.
1898.			
May 9	twice	3·412	2·000
,, 10	,,	3·046	1·864
,, 11	,,	6·304	3·812
,, 12	,,	2·381	1·547
,, 13	,,	2·262	1·190
,, 14	once	5·000	2·619
,, 16	twice	2·941	1·529
,, 17	,,	4·353	2·588
,, 18	,,	3·224	1·765
,, 19	,,	6·941	3·529
,, 20	,,	3·176	1·412
,, 21	once	6·000	2·706
,, 23	twice	4·128	1·517
,, 24	,,	3·042	1·639
,, 25	,,	6·934	2·552
,, 26	,,	2·170	1·012
,, 27	,,	3·544	1·531
Rest for Whitson Holidays.			
June 1	twice	3·725	1·126
,, 2	,,	4·010	1·549
,, 3	,,	3·091	1·440
,, 4	once	3·357	1·238
,, 6	twice	4·052	1·549
,, 7	,,	2·966	1·420
,, 8	,,	5·701	2·013
,, 9	,,	3·433	1·119
,, 10	,,	3·327	1·027
,, 11	once	2·094	·845
,, 13	twice	3·218	1·010
,, 14	,,	2·195	·857
,, 15	,,	6·340	2·314
,, 16	,,	4·041	1·229
,, 17	,,	2·764	1·324
,, 18	once	3·333	1·018
,, 20	twice	6·445	3·950
,, 21	,,	4·213	2·194
,, 22	,,	5·000	2·745
,, 23	once	7·412	4·926
Rest from June 23rd to July 10th.			

SINGLE AND DOUBLE COKE-BEDS.

Date	Number of grains of oxygen absorbed in 4 hours by one gallon of			
	Raw sewage.	Effluent from the single coke-bed (4 feet).	Effluent from the primary coke-bed (6 feet).	Effluent from the secondary coke-bed (6 feet).
1898.				
July 12	2·948	1·684	1·684	·842
,, 13	3·438	1·458	1·770	·833
,, 14	3·473	1·788	2·000	1·052
,, 15	2·000	1·472	1·684	·631
,, 16	3·474	1·472	1·800	·631
,, 18	1·702	·850	·957	·532
,, 19	6·702	2·553	3·085	1·808
,, 20	3·085	1·808	1·480	·850
,, 21	4·213	1·808	1·916	1·143
,, 22	2·065	1·260	1·260	·644
,, 25	1·950	1·055	·975	·529
,, 26	1·961	1·168	1·382	·529
,, 27	3·307	1·371	1·653	·952
,, 28	2·584	1·069	1·395	·847
,, 29	7·610	2·314	2·647	1·666
Rest to August 2nd (Bank Holiday).				

TABLE I.—SINGLE AND DOUBLE COKE-BEDS—(continued).

Date.	Number of grains of oxygen absorbed in 4 hours by one gallon of			
	Raw sewage.	Effluent from the single coke-bed (4 feet).	Effluent from the primary coke-bed (6 feet).	Effluent from the secondary coke-bed (6 feet).
1898.				
August 2	4·000	1·900	2·400	1·100
" 3	2·500	1·500	1·600	·700
" 4	3·131	1·717	1·515	·920
Rest on 5th and 6th while the spindle of the pump was being repaired.				
August 8	2·166	1·260	1·374	·683
" 9	3·955	1·636	1·450	·802
" 10	3·333	1·343	1·343	·644
" 11	4·207	1·640	1·759	1·077
" 12	2·641	1·295	1·295	·647
" 13	1·958	1·052	1·295	·490
" 15	2·254	1·052	1·176	·355
" 16	3·211	1·177	1·400	·818
" 17	3·750	1·873	1·666	·737
" 18	3·333	1·666	1·666	1·010
" 19	2·580	1·400	1·284	·752
" 20	2·010	·982	1·006	·642
" 22	3·933	1·641	1·422	·834
" 23	2·746	1·112	1·317	·834
" 24	4·211	2·000	2·000	1·196
" 25	2·263	1·118	1·010	·660
" 26	3·044	1·236	1·010	·749
" 27	4·400	1·717	1·994	·997
" 29	2·128	·957	1·383	·745
" 30	5·319	2·234	2·340	1·064
" 31	6·809	1·383	1·702	·851
September 1	6·145	3·229		
" 2	3·617	1·064	} Stopped working	
" 3	4·574	1·596		
" 5	3·191	1·489	1·596	1·064
" 6	5·417	2·187	2·708	1·362
" 8	4·632	2·211	2·316	1·684
" 9	3·152	1·196	1·848	·978
" 10	4·516	·860	1·628	1·505
" 12	4·193	1·505	2·043	1·505
" 13	4·946	2·903	2·258	1·398
" 14	5·319	1·915	2·447	1·596
" 15	3·830	1·596	1·915	1·170
" 16	6·774	3·548	4·086	3·118
" 17	3·298	1·277	1·596	1·064
" 19	2·234	1·277	1·809	1·170
" 20	5·745	1·809	2·234	1·702
" 21	4·574	2·128	2·128	1·489
" 22	4·239	2·174	1·957	1·304
" 23	6·129	1·935	3·222	2·258
" 24	4·444	1·778	1·889	·889
" 26	2·258	1·183	1·398	1·290
" 27	3·936	2·128	2·234	1·489
" 28	2·766	No sample	1·915	1·170
" 29	2·447	1·489	1·277	1·064
" 30	3·617	2·340	2·447	1·809
October 1	3·656	2·558	2·688	1·290
" 3	3·391	1·522	1·304	·978
" 4	3·898	1·809	2·021	1·383
" 5	3·643	1·848	2·391	1·522
" 6	4·255	2·234	2·660	1·702
" 7	3·978	2·258	2·366	1·505
" 8	3·991	2·340	2·340	1·596
" 10	3·391	1·739	1·739	1·196
" 11	2·447	1·277	1·609	1·170
" 12	3·587	1·530	1·820	1·196
" 13	3·763	2·151	2·581	1·720
" 14	2·717	1·413	1·957	1·304
" 15	3·696	1·729	2·065	1·304
" 17	3·696	1·522	2·009	1·196
" 18	3·614	1·932	1·932	1·136
" 19	2·909	1·682	1·909	1·341
" 20	3·522	1·870	2·234	1·596
" 21	4·830	2·043	2·903	1·505
" 22	4·146	2·312	2·312	1·525

TABLE I.—SINGLE AND DOUBLE COKE-BEDS—(continued).

Date.	Number of grains of oxygen absorbed in 4 hours by one gallon of			
	Raw sewage.	Effluent from the single coke-bed (4 feet).	Effluent from the primary coke-bed (6 feet).	Effluent from the secondary coke-bed (6 feet).
1898.				
October 24	3·000	1·474	1·368	1·158
,, 25	2·401	1·263	1·570	1·150
,, 26	3·460	No sample	2·098	1·260
,, 27	3·578	1·580	1·684	·947
,, 28	2·632	1·369	1·684	·947
,, 29	3·474	1·158	1·789	·842
,, 31	3·895	·737	·947	·526
November 1	2·948	·842	1·158	·737
,, 2	4·315	1·158	1·263	·842
,, 3	4·842	1·369	1·369	·737
,, 4	4·526	1·789	2·000	1·263
,, 5	5·148	1·790	1·580	1·158
,, 7	3·333	1·183	1·075	·860
,, 8	4·839	1·505	1·505	·968
,, 9	6·666	1·936	2·258	1·505
,, 10	4·839	1·936	2·043	1·183
,, 11	4·839	2·258	2·581	1·390
,, 12	5·159	2·366	2·150	1·505
,, 14	3·646	1·770	1·980	1·354
,, 15	4·167	2·188	2·083	1·770
,, 16	2·917	1·354	1·354	·938
,, 17	4·686	2·083	2·400	1·458
,, 18	3·437	1·770	1·875	1·354
,, 19	4·800	2·397	2·917	2·083
,, 21	3·510	1·596	1·496	·958
,, 22	2·979	1·702	1·596	1·064
,, 23	4·255	1·947	2·280	1·386
,, 24	3·617	1·915	1·809	1·170
,, 25	4·149	1·702	2·021	1·170
,, 26	2·766	1·439	1·702	·851
,, 28	4·639	2·211	2·315	1·702
,, 29	3·750	2·000	1·902	1·146
,, 30	3·646	1·042	1·875	·968
December 1	3·298	1·657	1·667	1·354
,, 2	2·708	1·224	1·354	·842
,, 3	2·812	1·354	1·579	1·170
,, 5	2·948	1·580	1·580	1·052
,, 6	3·789	1·684	1·895	1·370
,, 7	2·234	1·277	1·170	·851
,, 8	4·255	2·128	2·021	1·170
,, 9	3·510	2·021	1·809	1·064
,, 12	2·234	1·370	1·064	·851
,, 13	3·043	1·290	1·489	·957
,, 14	3·778	1·915	2·021	1·489
,, 15	2·979	1·595	1·595	1·070
,, 16	4·374	2·447	2·340	1·277
,, 17	2·842	1·489	1·489	·959
,, 19	3·750	2·083	2·083	1·354
,, 20	4·628	2·400	2·701	1·875
,, 21	2·500	1·500	1·354	·833
,, 22	3·125	1·701	1·701	·938
,, 23	3·958	1·876	2·500	1·701
,, 28	3·474	1·791	1·895	1·052
,, 29	2·948	1·473	1·791	·947
,, 30	5·155	3·093	2·800	1·443
,, 31	3·607	1·752	1·866	1·134
1899.				
January 2	3·100	1·500	1·600	1·000
,, 3	3·263	1·684	1·684	1·172
,, 4	3·548	1·915	1·753	·968
,, 5	2·553	1·383	1·250	·851
,, 6	4·255	2·426	2·234	1·489
,, 7	2·632	1·489	1·489	1·172
,, 9	2·653	1·328	1·429	1·124
,, 10	4·082	2·245	2·449	1·430
,, 11	3·535	1·818	1·818	1·111
,, 12	3·300	1·700	1·500	1·000
,, 13	2·600	1·300	1·600	·800
,, 14	3·800	2·000	2·100	1·400
,, 16	3·918	2·268	2·062	1·237

TABLE I.—SINGLE AND DOUBLE COKE-BEDS—(continued).

Date.				Number of grains of oxygen absorbed in 4 hours by one gallon of			
				Raw sewage.	Effluent from the single coke-bed (4 feet).	Effluent from the primary coke-bed (6 feet).	Effluent from the secondary coke-bed (6 feet).
1899.							
January 17	...	...	...	2·680	1·444	1·546	1·237
,, 18	...	...	...	3·390	1·939	1·633	1·020
,, 19	...	...	...	3·082	1·531	1·663	·918
,, 20	...	...	...	4·286	2·450	2·551	1·429
,, 21	...	...	...	4·000	1·895	2·105	1·158
,, 23	...	...	...	2·800	1·500	1·600	·900
,, 24	...	...	...	3·232	1·717	1·717	1·111
,, 25	...	...	...	3·030	1·717	1·616	·888
,, 26	...	...	...	2·500	1·800	1·500	1·000
,, 27	...	...	...	3·900	1·900	2·200	1·100
,, 28	...	...	...	3·417	1·935	1·709	1·093
,, 30	...	...	...	3·093	1·752	1·752	1·031
,, 31	...	...	...	2·680	1·546	1·444	·928
February 1	...	...	...	4·000	2·152	2·316	1·052
,, 2	...	...	...	2·959	1·633	1·735	·817
,, 3	...	...	...	4·286	2·450	2·551	1·428
,, 4	...	...	...	3·542	1·980	1·771	1·146
,, 6	...	...	...	3·480	1·826	1·938	·965
,, 7	...	...	...	2·844	1·630	1·630	·933
,, 9	...	...	...	4·275	2·200	2·436	1·112
,, 10	...	...	...	3·500	1·800	1·900	1·000
,, 11	...	...	...	4·061	2·354	2·579	1·270
,, 13	...	...	...	3·488	1·660	1·959	1·163
,, 14	...	...	...	4·048	2·326	2·471	1·294
,, 15	...	...	...	3·372	1·647	1·895	1·170
,, 16	...	...	...	2·706	1·512	1·512	·948
,, 17	...	...	...	2·495	1·163	1·368	·841
,, 18	...	...	...	3·139	1·647	1·774	1·059
Average	...	...	...	3·696	1·747	1·851	1·137

DIVISION II.—BACTERIOLOGICAL.

BY

A. C. HOUSTON, M.B., D.Sc.

CONTENTS.

A.—INTRODUCTION.

B.—SUMMARY OF CONTENTS OF FIRST REPORT.

I.—THE BIOLOGICAL TREATMENT OF SEWAGE.

II.—GENERAL RESULTS OBTAINED (MAY 9TH TO AUGUST 9TH, 1898).*

III.—SUMMARY OF RESULTS SHEWN IN TABLE 1.
  1. Total number of Bacteria.
  2. Number of Spores of Bacteria.
  3. Number of Liquefying Bacteria.
  4. Species of Micro-organisms.
      (a) B. Enteritidis Sporogenes (Klein).
      (b) B. Coli Communis.
      (c) Other species of Bacteria.

IV.—TABLES AND DIAGRAMS DEALING WITH THE RESULTS OF THE BACTERIOLOGICAL EXAMINATION OF THE CRUDE SEWAGE, OF THE EFFLUENTS FROM THE COKE-BEDS, OF THE EFFLUENT FROM THE CHEMICAL PRECIPITATION WORKS, AND OF SAMPLES OF THAMES WATER.

(The Diagrams will be found following page 41 of the text.)

V.—DESCRIPTION OF SOME OF THE BACTERIA FOUND IN THE CRUDE SEWAGE, AND IN THE EFFLUENTS FROM THE COKE-BEDS.
  1. B. Coli Communis.
  2. B. Mesentericus.
      Sewage variety E.
      Sewage variety I.
  3. Sewage Proteus.
  4. B. Frondosus.
  5. B. Fusiformis.
  6. B. Subtilissimus.
  7. B. Subtilis.
      Sewage variety A.
      Sewage variety B.
  8. B. Membraneus Patulus.
  9. B. Capillareus.

VI.—DESCRIPTION OF MICRO-PHOTOGRAPHS AND DIAGRAMMATIC DRAWINGS ACCOMPANYING THE REPORT.†

VII.—ADDENDA A, B, C, D, E.—FURTHER BACTERIOLOGICAL RECORDS FROM AUGUST 9TH TO DECEMBER 31ST, 1898.

(These Addenda follow Diagram 9 at the end of the Report.)

---

* Since this report was written (August 9th, 1898) circumstances arose delaying its publication for several months. To have brought it up to date would have necessitated rewriting the whole report and reconstructing all the Diagrams and Tables. Under these circumstances it was considered best to incorporate some of the later results in the form of addenda (Add. A, B, C, D, E), so that anyone interested in the progress of the coke-beds might follow their history a stage further. All the results, however, are not given, and it is hoped that in the future an opportunity may present itself of recording the work done since August 9th, 1898, in a more complete form.

† The micro-photographs were specially taken by Dr. Albert Norman. I wish to record my thanks to him for the great skill which he has shown in photographing my cultures and microscopic preparations, both for this and for my previous report.—A. C. H.

DIVISION II.—BACTERIOLOGICAL.

RESULTS OF THE BACTERIOLOGICAL EXAMINATION OF CROSSNESS CRUDE SEWAGE AND OF THE EFFLUENTS FROM THE COKE-BEDS.

A.—INTRODUCTION.

It has been stated (Division I., Section I., Sub-section 2A) that one of the main objects of the experiments was to ascertain—

"The effect of the treatment on the number and nature of the bacteria which were present in the raw sewage."

This portion of the Report contains a record of the work which has been done in this direction from May 9th to August 9th, 1898.[*]

It will be readily understood that, notwithstanding the large amount of work which has been carried out in obtaining data for this preliminary statement, the period during which the coke-beds have been in operation is too short to allow of anything in the nature of a final opinion being given as to their efficiency from the view-point of the bacteriologist. What follows then must be regarded as in a sense provisional, although it is probable that many of the results which have been obtained, and some, at all events, of the conclusions that have been arrived at will hold good in future work.

In what follows reference will be repeatedly made to the effluent from the 4-foot coke-bed, to the effluent from the 6-foot coke-bed, and also to the effluent from the laboratory coke-vessel. This last is the effluent from the 6-foot coke-bed which has again been treated in the small laboratory vessel at Crossness. A description of the construction and history of these coke-beds, together with all other essential particulars relating to them, will be found in Division I. of this Report (pages 7 to 9).

The Committee will remember that they have already received a report dealing with the bacteriology of London crude sewage as it is delivered at the Barking and Crossness Outfall Works.[†] Before proceeding further it is desirable to give a brief summary of the contents of this first report.

B.—SUMMARY OF CONTENTS OF THE FIRST REPORT.

Under the heading "Description of some of the Methods used in the Bacteriological Examination of Sewage," information was given as to (1) collection of samples; (2) dilution of sewage; the best way of estimating (3) the total number of bacteria; (4) the number of spores of bacteria; (5) the number of liquefying bacteria; and how to search for (6) special micro-organisms such as *B. coli*, typhoid bacillus, diphtheria bacillus, staphylococci and streptococci, and *B. enteritidis sporogenes*.

Under the heading "Summary of Results shown in Table I.," it was shown that (1) the total number of bacteria in 1 c.c. of Barking crude sewage averaged 3,899,259 (19 cultures of 9 samples), and in 1 c.c. of Crossness crude sewage 3,526,667 (11 cultures of 6 samples); (2) the number of spores of bacteria in 1 c.c. of Barking crude sewage averaged 332 (excluding certain extreme results), and in 1 c.c. of Crossness crude sewage 365 (excluding extreme results), which gives a ratio of spores to total number of bacteria of 1 to 11,744 and 1 to 9,562 respectively; (3) the number of liquefying bacteria in 1 c.c. of Barking crude sewage averaged 430,730, and in 1 c.c. of Crossness crude sewage 400,000, which gives a ratio of liquefying bacteria to total number of bacteria of 1 to 9 and 1 to 88 respectively. As regards (6) the species of micro-organisms, it was shown that in Barking and Crossness crude sewage the spores of *B. enteritidis sporogenes* were present in numbers varying from at least 10 to about 1,000 per c.c., and *B. coli* was present in numbers usually exceeding 100,000 per c.c. Notes were also given of the occurrence of other bacteria such as proteus-like forms, *bacillus fluorescens liquefaciens* and *non-liquefaciens B. mycoides*, *B. mesentericus*, *B. subtilis*, &c.

In Table I. were given the results of the bacteriological examination of 9 samples of Barking and 6 samples of Crossness crude sewage as regards the total number of bacteria, the number of spores of bacteria and the number of liquefying bacteria, and as regards the species of micro-organisms present.

A number of micro-photographs illustrating the work accompanied the Report.

It was said (page 1, A.—First Report) that the Report was only to be regarded as a preliminary statement of the progress of the work. By this it was not meant that the amount of work that had been carried out was of an insignificant character, the contrary was the case, but the subject was one of such a complex nature, and one about which so little is known, that it might well occupy the energies of not one but many workers for a very considerable period of time. Had the work merely covered the ground taken up by earlier observers it would still have been a most desirable thing from a scientific point of view to have confirmed or modified the results obtained in the past, particularly in an enquiry dealing with a subject which is still in its infancy, namely, the bacteriology of crude sewage.

But to a very considerable extent the research covered new ground, as, for example, the determination not only of the presence but of the relative abundance of a number of different micro-organisms, but more particularly of the *B. coli* and the spores of *B. enteritidis sporogenes*. *B. coli* is an example of an aerobe peculiarly abundant in crude sewage, and *B. enteritidis sporogenes* is an example of a pathogenic anaërobe less abundant, but perhaps more characteristic of sewage. The importance of obtaining records of the number of these germs in raw sewage preparatory to a study of their numbers in the effluents from the coke-beds needs no comment to show its importance. Moreover, such records are highly important from the point of view of the bacterioscopic analysis of drinking water.

---

[*] At the end of the report will be found further bacteriological records (9th August to 31st December, 1898), in addenda A, B, C, D, E.

[†] Filtration of sewage. Report on the bacteriological examination of London crude sewage. First Report (P. S. King and Son, 2 and 4, Great Smith-street, Westminster, S.W.).

[3]

I.—THE BIOLOGICAL TREATMENT OF SEWAGE.

The term biological, as applied to the treatment of sewage by natural processes, although manifestly convenient, is apt to be misleading. It is meant to denote the endeavour to obtain, under control, results similar to those which are constantly being produced in nature, and to employ processes which have been in operation for unknown ages. The term, however, is misleading because it might give the erroneous impression that sewage does not contain in itself the living organisms which are necessary to effect its decomposition, and that some new discovery had been made by which sewage was artificially treated with foreign bacteria, in order to bring about its destruction. This, however, is not the case.

Sewage already contains all the organisms which are necessary for its decomposition and final purification. In order to discover a method of biological treatment all that has to be done is to discover the best and the most practical way of allowing the natural purification by the action of the bacteria to take place without nuisance or danger.

In brief, the so-called biological treatment of sewage is neither more nor less than the attempt to imitate nature's own methods of effecting the decomposition and finally the purification of the effete matter of the animal and vegetable kingdom.

In this connection the testimony of Duclaux may be quoted:—"Whenever and wherever there is a decomposition of organic matter, whether it be the case of a herb or an oak, of a worm or a whale, the work is exclusively done by infinitely small organisms. They are the important, almost the only, agents of universal hygiene; they clear away more quickly than the dogs of Constantinople or the wild beasts of the desert the remains of all that has had life; they protect the living against the dead. They do more: if there are still living beings, if, since the hundreds of centuries the world has been inhabited, life continues, it is to them we owe it."

To describe the steps which have brought our knowledge to such a point that sanitarians now dare to look forward to a possible solution of the problem of sewage disposal would be to outline the history of bacteriology. Here it is permissible only to indicate in a few sentences some of the discoveries which have paved the way towards a scientific knowledge of the true nature of putrefactive processes.

One of the most important discoveries of recent times was that made by Schlœsing and Muntz in 1877. They proved that nitrification, or the oxidation of ammonia to nitric acid, is due to the vital activity of bacteria, and thus carried a stage further the important discovery by Schwann and Schultze (1839) that micro-organisms are the true agents of decomposition. Later, Winogradsky and others described and isolated, in pure culture, nitrifying organisms.

Pasteur, following up the researches of Cagniard and Schwann, demonstrated, in 1857, the relation between lactic, acetic, and butyric fermentations and special organisms. Later, he erroneously assumed that putrefaction was brought about solely by anaërobic bacteria. Since 1857 the science of bacteriology has advanced by leaps and bounds, and increased knowledge of the nature of putrefactive processes has, during the last decade, led to the serious attempt to bring about putrefaction, under control, as a means of disposing of excremental matters.

Putrefaction has been called putrid fermentation, but putrefaction is rather to be looked upon as a fermentation which is putrid or otherwise, according to the conditions under which it is conducted and the degree of its completeness.

The products of decomposition by so-called putrefaction are of the most varied kind, and there is no doubt that putrefaction is brought about, not by a single germ, but by a large number of different bacteria, some aërobic, others anaërobic. Each one of these germs may produce intermediary products of widely different character, but all of them, perhaps, tend in the direction of finally resolving highly complex organic substances into their simplest component parts.

Although the study of fermentation and the proper understanding of the true nature of putrefactive processes are attended with great difficulties, the aim and object of the biological treatment of sewage is sufficiently plain and simple.

In nature the following cycle of transformation takes place:—Dead organic matter decays as the result of the vital activity of bacteria, and ammonia is liberated. The nitrifying organisms bring about the oxidation of the ammonia, first to nitrous and then to nitric acid. These acids by reaction upon the bases, always present, form nitrites and nitrates, and these nourish the living plant. While the nitrogen is undergoing these changes, the carbon of the organic matter is converted into carbonic acid, and the hydrogen mainly into water. To some extent also the nitrogen and hydrogen are liberated in the free gaseous state.

Although organic matter is ultimately resolved into innocuous, inorganic substances by these processes, the intermediary products of decomposition, as has been already indicated, may be of a highly complex character. Fortunately, however, these complex substances are for the most part unstable in character and readily oxidisable, and from the practical point of view they may be regarded therefore as the earlier stages leading up to the final purification.

Now the organic matters found in sewage are partly in suspension and partly in solution, and sewage contains in itself the necessary living germs for the destruction of both these forms of organic matter. The aim and object of the biological treatment of sewage is to render soluble by microbial agencies the solid matters; and to split up by the action of living bacteria both the matter thus dissolved and the organic compounds which were originally in solution, into their simpler elements. In the final process of purification, these substances should undergo oxidation induced by the life processes of nitrifying organisms, and an effluent should be produced which is free from putrescible matter and contains only inorganic or mineral substances.

Of course it will be argued that there is great danger in encouraging the life processes of micro-organisms which are largely derived from the excreta of animals. Such a contention is certainly justifiable. But in this connection it is to be noted in the first place that the biological treatment of sewage is conducted *under control*; secondly, that the process always gradually secures the destruction of the pabulum on which bacteria feed, and hence leads to their death; thirdly, that the balance of evidence points to the probability that some, at all events, of the pathogenic organisms are crowded out in the struggle for existence in a nutritive medium containing a mixed bacterial flora,* their vitality being weakened or destroyed by the enzymes of the saprophytic species;† fourthly, that while it is true that bacteria produce poisonous substances in their growth, it also is true that their chemical poisons are toxic in proportion to the dose, and, moreover, are highly unstable and readily break down into their elementary and innocuous constituents; and, lastly, that in some cases it may not be necessary to attempt the complete purification of the sewage, the solution of the suspended matters and partial destruction of the putrescible matter in solution being all that is urgently called for, as, for example, where the effluent is of relatively small bulk and is turned into a stream the water of which is not used for domestic purposes (as is the case in the lower Thames) or else when the effluent is to be subsequently treated by land irrigation.

It has been said that the processes of putrefaction and nitrification are dependent on the vital activity of bacteria, and that the destruction of the organic matters in sewage can be effected by the life processes of living germs. But it will be asked under what conditions the bacteriological purification of sewage can be best carried out, and whether there is any process known which in point of economy, practicability and freedom from obvious danger compares with the methods ordinarily in use—namely, chemical treatment and land irrigation. These are questions which cannot be fully answered in the present state of our knowledge.

As regards the conditions essential to the successful biological treatment of sewage, it may be said at the outset that the presence in the sewage of any substance inimical to microbial life must, if possible, be guarded against. Thus, it is well known that acidity inhibits the growth of bacteria, and that slight alkalinity is a favourable condition. Again, many of the waste products of manufacturing processes are poisonous to bacteria, and in the preliminary treatment of the sewage with certain chemicals it may also be rendered unfit for bacterial life. But whether the sewage is best placed primarily under aërobic or under anaërobic conditions is a question still open to discussion. All, however, are agreed that the final treatment must be aërobic.

As regards a comparison between the bacteriological treatment of sewage and its treatment by chemical precipitation or by land irrigation, it may safely be said that even the most powerful advocates of treatment with chemicals or treatment on land are dissatisfied with the general results that have been obtained in the past, and it is a significant fact that the believers in the possibility of the so-called self-purification of sewage are rapidly increasing in numbers. Moreover, the records, which are daily accumulating, of the practical trials of bacterial coke-beds in different parts of the country tend in the direction of demonstrating the possibility of purifying sewage biologically, although the conditions under which such treatment is best carried out are not yet fully known.

One point which has been rather lost sight of is that there is probably no bacterial process in practical operation at the present time which is not eminently successful in that it places the sewage in a most favourable state for its final purification by land irrigation, or by other methods. The mere solution of the great mass of the suspended matters by bacterial agencies, which is, perhaps, common to all the different processes at present under trial, is a sufficient vindication of the enormous advantage to be gained by the biological treatment of sewage. And, even when an effluent is excessively foul, it is, perhaps, not wholly justifiable to condemn it in *unmeasured terms*, because such foulness may possibly be a step, although not of the most direct character, leading towards ultimate purification.

The scope of this report permits only of a brief allusion to some of the methods which are in operation for dealing bacteriologically with raw sewage and partially purified effluents. No attempt, however, will be made to describe their respective merits.

To the State Board of Health of Massachusetts belongs the honour of having carried out a laborious series of experiments upon the purification of sewage at a time when the whole subject was enshrouded in mystery. The publication of the valuable results of this investigation at once aroused the interest of sanitarians in this country.

In 1892 Mr. Scott-Moncrieff introduced his cultivation bacteria-bed. He founded his system on the following well-recognised truths—

(1) That bacteria under favourable conditions are capable of indefinite multiplication.

(2) That there exist in sewage, bacteria which are capable of peptonizing solid organic matter, or in other words, of preparing it, by a process comparable to that of digestion, for its final disintegration.

(3) That in nature the purification of the refuse of the organic world is effected by the life history of these or similar micro-organisms.

In this system the crude sewage passes into the bottom of a bed filled with flint, coke and gravel, and fitted with a false bottom. The liquid portion rises through this bed and the suspended matters are held back at the bottom and undergo solution by the action of bacteria. In order to still further improve the quality of the effluent, by oxidation and by the action of micro-organisms, the effluent from this primary bed is afterwards passed down nitrifying channels. Since 1892 Mr. Moncrieff has modified his original plans in a number of important respects.

---

*In this connection reference may be made to certain interesting experiments carried out by Lawes and Andrewes, and published in their joint report to the London County Council on Micro-organisms in Sewage, 1894.

† It must be distinctly understood that I do not imply that such organisms as the typhoid bacillus or the cholera vibrio would necessarily lose their vitality, or even suffer a diminution in virulence under the conditions prevailing in a biological filter. In the absence of actual experiments with the particular sewage in question, I am not prepared to say more than that I believe that if these germs did gain access to the sewage they would suffer a diminution in numbers primarily in the sewers, and secondarily in the coke-beds.—A. C. H.

While Mr. Moncrieff was engaged in demonstrating the possibility of disposing of sewage by the agency of bacteria, other workers appeared in the field. In particular, Messrs. Adeney and Parry commenced a long series of investigations which has finally resulted in the formation of a company called The Oxygen Sewage Purification Company. This so-called oxygen system of sewage purification possesses three main features of interest, namely—

 I. The use of crude manganese compounds for the purpose of clarification or precipitation.
 II. The principle of recovering the chemicals employed for the precipitation.
 III. The use of nitrate of soda as a substitute for air (and therefore of the coke-bed) for supplying oxygen to the organisms in the manner required by them.

Mr. Cameron's system was brought forward in 1895 at Exeter. In this system the sewage is first passed into a closed tank with the object of producing liquefaction of the solid matters by anaërobic bacteria. The effluent from this anaërobic tank is next passed over an aërating weir and thence to a series of coke-beds. The arrangement is such that each bed in turn is filled, remains full for a period, is then emptied, and finally is allowed a rest before being again filled.

In 1897 Colonel Ducat introduced his "aërated bacterial self-acting coke-bed." This is a *continuous filtration process*, the raw sewage falling on the surface of the open bed, the walls of which are made of open drain pipes, and after passing through the body of the bed, which is composed of coarse material at the top and fine at the bottom, finally escapes without having been exposed to any anaërobic treatment.

It is unnecessary to describe the system of sewage disposal advocated by Messrs. Dibdin and Thudichum, because the Committee have from time to time received the records of their prolonged investigations.

Messrs. Waring and Lowcock, although working on different lines, attempt to bring about the oxidation of the organic matter in their beds by pumping operations.

At Davyhulme, near Manchester, an interesting series of experiments have been carried out under the direction of Sir Henry Roscoe with coke, cinder, coal, and sand beds, the treatment of the effluent from the sewage precipitation tanks being in some cases continuous and in others intermittent.

II.—GENERAL RESULTS OBTAINED AT THE CROSSNESS OUTFALL
(May 9th to Aug. 9th, 1898*).

Under the above heading are given the chief results which have been obtained from the effluent of the experimental 4-foot coke-bed. The records of the examination of the effluent from the other beds are too small in number to allow of their being included here, but reference is made to them in the body of the report.

TOTAL NUMBER OF BACTERIA PER C.C.

Crude sewage   6,140,000 (average of 10 experiments)
4-foot coke-bed effluent   4,437,500 (average of 8 experiments)
Percentage reduction   27·7

NUMBER OF SPORES OF BACTERIA PER C.C.

Crude sewage   407 (average of 10 experiments)
4-foot coke-bed effluent   252 (average of 8 experiments)
Percentage reduction   38

LIQUEFYING BACTERIA PER C.C.

Crude sewage   860,000 (average of 10 experiments)
4-foot coke-bed effluent   762,500 (average of 8 experiments)
Percentage reduction   11·3

SPORES OF B. ENTERITIDIS PER C.C.

Crude sewage   From 10 to 1,000 (usually more than 100) (11 experiments)
4-foot coke-bed effluent   Do.   Do.   (10 experiments)
Percentage reduction   Practically no reduction

B. COLI PER C.C.

Crude sewage   More than 100,000 (10 experiments)
4-foot coke-bed effluent   Do.   (8 experiments)
Percentage reduction   Practically no reduction

MICRO-ORGANISMS OTHER THAN B. ENTERITIDIS AND B. COLI.

Crude sewage } In comparing the cultures no distinct difference could be made out, as regards
4-foot coke-beds effluent } the species of microbes, between the cultivations made from the crude sewage and those made from the effluents.

The records are not sufficiently extensive to allow of final deductions being drawn. Yet it may be worthy of note that the percentage reduction of spores of bacteria was greater, and the percentage reduction of liquefying bacteria less than the reduction of the total number of micro-organisms.

---

* Some of the later results (Aug. 9th to Dec. 31st, 1898), are given in Addenda A, B, C, D, E. It must however, be understood that the descriptive matter in this report deals only with the results up to Aug. 9th. Nevertheless, the general results obtained since then do not differ very widely from those here recorded.

It might reasonably be inferred that in the coke-beds there was a relative increase in the liquefying bacteria, and a relative decrease in the number of spores, as a result of the biological treatment of the sewage.

It must be admitted that the above results are not satisfactory from the bacteriological point of view, particularly when it is remembered that an effluent ought to be judged, not only by the percentage amount of purification effected, but also by the *actual state* it is in. Yet it is to be considered that these results assume a different complexion when viewed side by side with the chemical data.

It has been shown in Division I. of this report that the percentage purification, as judged by the dissolved oxidisable matter removed by the treatment, was on an average 51·3 per cent., and that the suspended matter was entirely removed. It has been stated that the results thus obtained surpass considerably those yielded by chemical treatment and appear to justify the claims put forward by the supporters of the biological treatment of sewage, especially since, so far as can be seen, no nuisance or danger arises as a result of the treatment.

In the body of the report a number of reasons are given, showing that it is unwise in the present state of our knowledge to recklessly condemn an effluent on bacteriological ground alone, without full knowledge of all the requirements of the case. In the attempt to treat sewage on biological lines it is to be noted that the solution of the suspended matter and even the partial destruction of putrescible matters by microbial agencies afford sufficient ground for justifying the process, at all events as a preliminary measure. Whether this preliminary treatment is to be supplemented by further treatment, either by passage through other coke-beds, or by land irrigation or by any other method, is a matter largely dependent on circumstances.

In the present case there are practical points which first of all demand consideration, and although it may be most desirable to obtain an effluent chemically pure and bacteriologically above suspicion of danger it is to be thought of that an effluent not altogether satisfactory in one or other, or even in both, of these respects may yet fulfil all necessary requirements without passing out of the range of practicability. In certain cases it may be imperative to obtain an effluent bacteriologically sound, but it does not follow that a similar result is urgently called for in other cases, as, for example, where an effluent is turned into a watercourse which is not used for drinking purposes, and which already may contain practically all the bacteria that are found in sewage.

It might reasonably be argued that where an effluent is turned into a river already grossly polluted and below the lowest level of "intake" for waterworks purposes, that the chemical state of such effluent was (from the practical point of view, at all events) of possibly even greater importance than the bacteriological. Some such state of things pertains in the case of London sewage and the river Thames. Here the initial consideration is to avoid fouling the river with putrescible matters to such an extent as to constitute a grave public nuisance. From this point of view it is evident that an effluent rich in putrescible matter is not permissible, but it is not as certain that an effluent rich in bacteria is equally to be condemned on practical grounds.

In conclusion, it must not be considered from what has been said that any attempt is made to minimise the importance of the bacteriological state of the effluent, or to disguise the true significance of the results that have been obtained. No doubt the question will eventually have to be faced—*Are the advantages gained by chemical purification sufficiently great to outweigh the possible danger arising from the discharge of an effluent bacteriologically unsound into the river Thames?*

That the water itself of the Thames is in an unsatisfactory state from the bacteriological point of view may be deduced from the results shown in Table 2. This table also shows that the effluents from the Barking and Crossness precipitation-tanks are no better, if indeed they are not worse, than average samples of the raw sewage.

III.—SUMMARY OF RESULTS* SHOWN IN TABLE I.

The results are those derived from the bacteriological examination of Crossness crude sewage, and of the effluents respectively from the 4-foot coke-bed, from the 6-foot coke-bed, and from the laboratory coke-vessel (see pages 28 to 31).

1.—Total number of bacteria in 1 c.c. (Table I., col. 2).

1898.	Crossness crude sewage.	Effluent from 4-foot bed.	Effluent from 6-foot bed.
May 11	3,930,000 (expt. 1)	4,800,000 (expt. 2)	... ...
,, 18	3,670,000 (,, 3)	4,100,000 (,, 4)	... ...
,, 25	6,400,000 (,, 5)	6,100,000 (,, 6)	... ...
June 9	6,500,000 (,, 7)	1,200,000 (,, 8)	... ...
,, 15	4,000,000 (,, 9)	5,300,000 (,, 10)	... ...
,, 22	9,100,000 (,, 11)	3,000,000 (,, 12)	... ...
July 20	12,800,000 (,, 15)	9,200,000 (,, 16)	... ...
,, 27	7,200,000 (,, 19)	... ...	6,600,000 (expt. 20)
August 4	4,200,000 (,, 23)	1,800,000 (,, 24)	... ...
,, 9	3,600,000 (,, 27)	... ...	1,700,000 (,, 28)
Highest number	12,800,000 (,, 15)	9,200,000 (,, 16)	6,600,000 (,, 20)
Lowest ,,	3,600,000 (,, 27)	1,200,000 (,, 8)	1,700,000 (,, 28)
Average ,,	6,140,000 (av. of 10 expts.)	4,437,500 (av. of 8 expts.)	4,150,000 (av. of 2 expts.)

These results are shown in graphic form in Diagram 3.

* I desire to record my thanks to Dr. Klein for advice and help of the greatest value.—A. C. H.

It will be noted that in these comparative experiments the total number of bacteria in the crude sewage usually exceeded those found in the effluents from the coke-beds. On three separate occasions, however, the effluent from the 4-foot coke-bed contained more organisms than the crude sewage, namely, expts. 2, 4, and 10, as compared with expts. 1, 3, and 9. In round numbers the averages were over six million, over four million, and over four million in the crude sewage, the effluent from the 4-foot bed, and the effluent from the 6-foot bed respectively. The percentage reduction of bacteria was on an average 27·7 in the case of the 4-foot bed effluent and 32·4 in the case of the 6-foot coke-bed effluent.

Although the percentage reduction in the total number of bacteria is fairly satisfactory, it must be conceded that the average number of microbes left in the effluent is very large, namely, 4,437,500 in the case of the 4-foot coke-bed, and 4,150,000 in the case of the 6-foot coke-bed.

It is probable that the conditions prevailing in the coke-beds may have been satisfactory as regards the solution of suspended organic matter and even as regards the partial destruction of dissolved organic matter, and yet have fallen short of leading to active nitrification and total purification. In the Massachussets experiments, when the reduction in the amount of organic matter was satisfactory and nitrification was in active progress, the total number of bacteria suffered a marked diminution.

In brief, it may be, that the coke-beds were highly efficient from the point of view of solution of suspended organic matter and partial destruction of dissolved and offensive substances, and as a preliminary measure tending in the direction of complete disintegration and ultimate purification of the crude sewage, but that they were unable to bring about the complete resolution of the organic matter; if this were the case, sufficient pabulum may have been left in the effluent to allow of the continued growth and multiplication of the sewage bacteria.

Indeed, so long as organic matter in an assimilable form remains in the effluent, multiplication of existing micro-organisms may be expected to take place until the self injurious products of these bacteria, or a lack of nutritive material, or some other conditions adverse to microbial life lead to their destruction or to a cessation of their powers of multiplication.

The presence of bacteria in enormous numbers in an effluent does not perhaps necessarily imply that the effluent is of a degraded character and highly putrescible; it may only mean that the liquid has passed through a previous stage of putrefaction, preparatory to its purification, in which case the danger so far as nuisance is concerned may be regarded as potential and not actual. It is as certain on the one hand that the addition of an infinite number of bacteria to a pure liquid, would be followed by a decrease in their number owing to the lack of nutritive material, as it is on the other hand that the removal of all the bacteria from a foul liquid would be a useless precaution, since Nature has always at her command a host of micro-organisms ready to attack the effete matter of the vegetable and animal kingdom, and effect directly or indirectly its purification.

Nevertheless, a liquid swarming with living bacteria is usually a liquid still undergoing putrefaction, and is likely also to contain germs of a harmful nature.

It may be worthy of note that out of the sixteen samples of Crossness and the nine samples of Barking crude sewage examined during a period extending from February to August, 1898, on no occasion was it found that the total number of bacteria was other than very large. It is evident then that the various substances, such as waste matters from manufactories, which are discharged in large quantities into the London sewers, and yet may be regarded as foreign to sewage, do not inhibit the growth of the sewage bacteria to any marked extent. It is important to note this fact, because if the bacterial purification of sewage should be adopted on a large scale, disastrous results might arise if the sewage was at one time rich in microbial life and at another almost sterile owing to the presence of foreign substances inimical to bacteria.

In conclusion, it must be borne in mind that the coke-beds at Crossness were purposely constructed so as not to lead to mere mechanical filtration of the suspended matter, much less of the bacteria in the crude sewage.

In estimating the total number* of bacteria gelatine plates were used. From 0·1 to 1·0 c.c. of crude sewage or effluent diluted with 10,000 times its volume of sterile water (i.e., 0·00001 to 0·0001 c.c. of the original fluid) was added to 10 c.c. of sterile gelatine contained in a test tube. After the gelatine had been melted, it was poured into a Petri's capsule, and after solidification had taken place the plate was inverted, incubated at 20 C., and the colonies subsequently counted at as late a date as the liquefaction of the gelatine and the crowding of the colonies allowed of.

2.—NUMBER OF SPORES OF BACTERIA IN 1 c.c. (TABLE I., COL. 3).

1898.	Crossness crude sewage.	Effluent from 4-ft. coke-bed.	Effluent from 6-ft. coke-bed.
May 11	460 (expt. 1)	260 (expt. 2)	...
„ 18	300 („ 3)	140 („ 4)	...
„ 25	370 („ 5)	380 („ 6)	...
June 9	560 („ 7)	230 („ 8)	...
„ 15	180 („ 9)	300 („ 10)	...
„ 22	310 („ 11)	60 („ 12)	...
July 20	400 („ 15)	430 („ 16)	...
„ 27	870 („ 19)	...	480 (expt. 20)
August 4	280 („ 23)	220 („ 24)	...
„ 9	340 („ 27)	...	300 („ 28)
Highest number	870 („ 19)	430 („ 16)	480 („ 20)
Lowest number	180 („ 9)	60 („ 12)	300 („ 28)
Average	407 (av. of 10 expts.)	252 (av. of 8 expts.)	390 (av. of 2 expts.)

These results are shown in graphic form in Diagram 4.

* See B 3, page 2—First Report.

In these comparative experiments the number of spores of bacteria in the crude sewage usually exceeded those found in the effluents from the coke-beds, but on three separate occasions the number in the effluent from the 4-foot coke-bed was greater than in the crude sewage—namely, experiments 6, 10, and 16 as compared with experiments 5, 9, and 15. The averages were 407, 252, and 390 in the crude sewage, in the effluent from the 4-foot bed and in the effluent from the 6-foot bed respectively. The percentage reduction of spores of bacteria was on an average 38 in the case of the 4-foot coke-bed effluent, and 4 in the case of the 6-foot coke-bed effluent. In the latter case, however, the average is based on two experiments only.

In the case of the 4-ft. coke-bed effluent, the percentage reduction in the number of spores was slightly greater than the percentage reduction of the total number of bacteria. Yet the number of spores remaining in the effluent, namely, 252 on an average, was large. Spores of bacteria are peculiarly resistant to unfavourable physical conditions; fortunately, however, the majority, at all events, of the spores of aërobic micro-organisms found in sewage belong to species which are believed to be harmless.

Taking the figures 6,140,000, 4,437,500 and 4,150,000 as representing the total number of bacteria, and 407, 252 and 390 as representing the number of spores in one cubic centimeter of an average sample of Crossness crude sewage, of effluent from 4-ft. bed, and of effluent from 6-ft. bed respectively, the ratios of spores to bacteria are as 1 to 15,086, 1 to 17,609, and 1 to 10,641 respectively. It appears, then, that there was a slight diminution in the number of spores, relative to the total number of bacteria, as a result of the treatment of the sewage in the 4-ft. coke-bed. The opposite is the case as regards the 6-ft. bed, but here only two samples of effluent were examined.

Comparing diagrams 3 and 4 as regards crude sewage, no direct parallelism appears to exist between the total number of germs and the number of spores, although to some extent a rise or fall in the total number of bacteria was associated with a rise or fall in the number of spores. When, however, a similar comparison is made in the case of the 4-ft. coke-bed effluent, it will be noted that a parallelism does exist between the total number and the number of spores of bacteria.

In estimating the number of spores of bacteria, * the following plan was adopted:—To 10 c.c. of sterile gelatine in a test tube was added 1 c.c. of diluted sewage or effluent (1:10), and the mixture heated to 80° C. for ten minutes and then poured into a Petri's capsule. After the gelatine had become quite solid the plate was inverted and incubated at 20° C.

3.—NUMBER OF LIQUEFYING BACTERIA IN 1 C.C. (TABLE I., COL. 4).

1898.	Crossness crude sewage.	Effluent from 4-ft. coke-bed.	Effluent from 6-ft. coke-bed.
May 11	400,000 (expt. 1)	1,300,000 (expt. 2)	...
" 18	100,000 (" 3)	700,000 (" 4)	...
" 25	900,000 (" 5)	700,000 (" 6)	...
June 9	1,600,000 (" 7)	200,000 (" 8)	...
" 15	800,000 (" 9)	1,100,000 (" 10)	...
" 22	900,000 (" 11)	600,000 (" 12)	...
July 20	1,700,000 (" 15)	1,000,000 (" 16)	...
" 27	900,000 (" 19)	...	300,000 (expt. 20)
August 4	400,000 (" 23)	500,000 (" 24)	...
" 9	1,100,000 (" 27)	...	200,000 (" 28)
Highest number	1,700,000 (" 15)	1,300,000 (" 2)	300,000 (" 20)
Lowest number	100,000 (" 3)	200,000 (" 8)	200,000 (" 28)
Average	860,000 (av. of 10 expts.)	762,500 (av. of 8 expts.)	250,000 (av. of 2 expts).

These results are shown in graphic form in Diagram 5.

Although the number of liquefying bacteria in the crude sewage largely exceeded those in the effluent from the 6-ft. coke-bed, the difference between the liquefying microbes in the 4-ft. coke-bed effluent and in the crude sewage was not great, and in four out of the eight comparative experiments the number was greatest in the former—namely, in experiments 2, 4, 10, and 24, as compared with experiments 1, 3, 9, and 23.

The averages were 860,000, 762,500, and 250,000 in the crude sewage, in the effluent from the 4-ft. bed, and in the effluent from the 6-ft. bed.

The percentage reduction was on an average 11·3 in the case of the 4-foot coke-bed effluent and 70·9 in the case of the 6-foot coke-bed effluent. But as regards the 6-foot coke-bed effluent, the average is based on the result of the examination of two samples only.

It is to be noted that in the case of the 4-foot coke-bed effluent the reduction in the liquefying bacteria, as compared with the reduction in the total number of microbes and in the spores of micro-organisms, was very small.

The ratio between the number of liquefying bacteria and the total number of bacteria in Crossness crude sewage, in the effluent from the 4-foot bed and the effluent from the 6-foot bed, is 1 to 7·1, 1 to 5·8, and 1 to 16·6 respectively.

As regards the 4-foot bed, it would seem as if the biological treatment of the sewage gave rise to a slight increase in the number of liquefying bacteria relative to the total number of germs. The opposite holds good in the case of the 6-foot bed, but here only two samples were examined.

Comparing diagrams 3 and 5 as regards the crude sewage it will be seen that a rise or fall in the total number of bacteria was to some extent associated with a rise or fall in the number of liquefying bacteria. But when in the case of the 4-foot coke-bed effluent a similar comparison is made, it will be noted that the parallelism between the two was more distinct.

Comparing diagrams 3, 4, and 5, it follows from what has been already said as regards the 4-ft. coke-bed effluent, that if the total number of bacteria and the number of spores, and the total number of bacteria and the number of liquefying organisms, are related in each case, then there must be some relation between

* See B 4, page 2—First Report.

the number of liquefying germs and the number of spores. Further, it has been said as regards the crude sewage, that no distinct parallelism exists between the total number of bacteria and the number of spores and the relation between the total number and the number of liquefying microbes is not well marked. Neither, on comparing diagrams 4 and 5, does there appear to be any distinct relation between the number of spores and the number of liquefying bacteria in the crude sewage.

In order to observe these facts more closely reference may be made to diagrams 6 and 7, which show the percentage deviation from the mean in the different samples of crude sewage and effluent from the 4-ft. and 6-ft. beds as regards total number of germs, number of spores, and number of liquefying bacteria.

Firstly, as regards the crude sewage, diagram 6 shows that there is no definite relation between the total number of bacteria and the number of spores. To some extent, however, there is a relation between the total number of bacteria and the liquefying germs.

Secondly, as regards the effluents from the coke-beds, diagram 7 shows that there is a distinct relation between the total number—the number of spores and the number of liquefying bacteria. Although, however, a rise or fall above or below the mean in the total number was nearly always coincident with a rise or fall above or below the mean in the number of spores and the number of liquefying bacteria, the percentage deviation in each case showed no parallelism.

Lastly, comparing diagrams 6 and 7, no relationship can be made out between the total number, the number of spores and the number of liquefying bacteria in the crude sewage, as compared with the effluent from the 4-ft. bed.

In the first report it was stated (page 4, c. 3) that—"It must not be concluded, however, that because a micro-organism liquefies gelatine, it will also liquefy all the various and complex kinds of organic matter existing in sewage. But it is safe to assert that a liquid rich in bacteria of varied species, many of which are capable of producing liquefaction of gelatine, is likely also to be rich in ability to dissolve solid or suspended organic matter."

As illustrating the complex behaviour of different germs to different albuminoids, and also of the same organism to different albuminoids, it may be noted that some bacteria liquefy gelatine, others do not; some coagulate the casein in milk and then dissolve it, others coagulate it and do not further peptonize it, others peptonize it directly ; again some organisms which liquefy gelatine coagulate milk, others coagulate milk but do not liquefy gelatine; again an organism which liquefies gelatine may or may not produce liquefaction of fibrin, of blood serum or of egg albumen, and an organism capable of peptonizing fibrin may fail to liquefy gelatine.

In a description of some of the bacteria found in the crude sewage and in the effluents which accompanies this report, will be found details relating to their behaviour when grown in milk, gelatine, blood serum, etc. For example, "sewage proteus" quickly liquefies gelatine and blood serum, and apparently peptonizes milk without first coagulating it. $B.$ $coli$ produces no liquefaction of gelatine or blood serum, but clots milk in 24 hours at 37°C. $B.$ $mesentericus$ $I.$ rapidly liquefies gelatine and blood serum, and apparently peptonizes milk without previous coagulation. $B.$ $mesentericus$ $E.$ liquefies gelatine very slowly, but liquefies blood serum fairly rapidly at 37°C., and produces a weak clot in milk which appears to be subsequently dissolved.

In estimating the number of liquefying organisms the following plan* was adopted—The contents of a test tube containing 10 c.c. of sterile nutrient gelatine were melted and poured into a sterile Petri's capsule. After the gelatine had become quite solid the surface of the medium was inoculated with 0·1 c.c. of diluted sewage or effluent (1 : 10,000). The diluted sewage or effluent (representing 0·00001 c.c.) was then spread over the entire surface of the gelatine with a platinum instrument. The plate was next inverted and incubated at 20° C. in this position until the colonies were sufficiently advanced in their growth for observation. Although this method is the best one available, it must be remembered that some bacteria liquefy the gelatine so very slowly that they might readily escape being counted as liquefying germs under the above conditions of experiment. This matters the less since bacteriologists are in the habit of classing some, at all events, of these bacteria as non-liquefiers.

4.—SPECIES OF MICRO-ORGANISMS PRESENT IN CROSSNESS CRUDE SEWAGE AND IN THE EFFLUENTS FROM THE COKE-BEDS.

(a) $Bacillus$ $enteritidis$ $sporogenes$ $(Klein).$†

In the First Report the result of a considerable number of experiments were given, showing that the spores of $B.$ $enteritidis$ may be present in London crude sewage‡ in numbers varying from 10 to 1,000 per c.c. Further, it was pointed out that Dr. Klein's researches tend to show that this organism is causally related to diarrhoea. That its cultures are extremely virulent may be seen by referring to col. 5, expt. 6, Table I., and col. 5, expts. 4 and 5, Table II., and also expts. 1 and 2, col. 5, Table I., of First Report. Again, it was stated that this pathogenic anaërobe is most important from the point of view of the bacterioscopic analysis of water, and lastly the hope was held out that future work would show what was the fate of $B.$ $enteritidis$ during the passage of the sewage through the biological coke-beds at the Outfall Works.

The following is a summary of the results shown in col. 5 of Table I.—

Date.	Crossness crude sewage.	Effluent from 4-foot coke-bed.	Effluent from 6-foot coke-bed.	Effluent from 6-foot coke-bed again passed through the laboratory vessel at Crossness.
1898.				
11th May	+ 0·1 c.c. — 0·01, 0·001 and 0·0001 c.c. sewage (expt. 1)	+ 0·01 c.c. — 0·001 and 0·0001 c.c. effluent (expt. 2)	...	...
18th May	+ 0·1, 0·01 and 0·001 c.c. sewage (expt. 3)	+ 0·1 and 0·01 c.c. — 0·001 c.c. effluent (expt. 4)	...	...

* See B 5, page 2—First Report.
† See figure 2, Plate I., of this report, and Nos. 3 and 14 (Plate I.), First Report.
‡ See page 5, and also Table I., of First Report.

Date.	Crossness crude sewage.	Effluent from 4-foot coke-bed.	Effluent from 6-foot coke-bed.	Effluent from 6-foot coke-bed again passed through the laboratory vessel at Crossness.
1898. 25th May	+ 0·1 and 0·01 c.c. − 0·001 c.c. sewage (expt. 5)	+ 0·1, 0·01 and 0·001 c.c. effluent (expt. 6)	...	...
9th June	+ 0·1 and 0·01 c.c. sewage (expt. 7)	+ 0·1 and 0·01 c.c. effluent (expt. 8)		
15th June	+ 0·1 c.c. sewage (expt. 9)	+ 0·1 c.c. effluent (expt. 10)		
22nd June	+ 0·1, 0·01 and 0·001 c.c. sewage (expt. 11)	+ 0·1 c.c. − 0·01 and 0·001 c.c. effluent (expt. 12)	...	...
6th July	+ 0·1 c.c. − 0·01 and 0·001 c.c. sewage (expt. 13)		+ 0·1 c.c. − 0·01 and 0·001 c.c. effluent (expt. 14)	...
20th July	+ 0·1 and 0·01 c.c. − 0·001 c.c. sewage (expt. 15)	+ 0·1 and 0·01 c.c. effluent (expt. 16)	+ 0·1 and 0·01 c.c. effluent (expt. 17)	+ 0·1 and 0·01 c.c. effluent (expt. 18)
27th July	+ 0·1 and 0·01 c.c. sewage (expt. 19)	+ 0·1 and 0·01 c.c. −0·001 c.c. effluent (expt. 21)	+ 0·1 and 0·01 c.c. effluent (expt. 20)	+ 1·0 and 0·1 c.c. − 0·01 c.c. effluent (expt. 22)
4th August	+ 0·1 and 0·01 c.c. sewage (expt. 23)	+ 0·1 and 0·01 c.c. effluent (expt. 24)	+ 0·1 c.c. effluent (expt. 25)	+ 0·1 c.c. effluent (expt. 26)
9th August	+ 0·1 c.c. − 0·01 c.c. sewage (expt. 27)	+ 0·1 and 0·01 c.c. effluent (expt. 29)	+ 0·1 and 0·01 c.c. effluent (expt. 28)	+ 0·1 c.c. − 0·01 c.c. effluent (expt. 30)

(The sign + signifies the presence, and the sign − the absence of the spores of *B. enteritidis sporogenes*.) The above results are shown in graphic form in Diagram 8.

As regards the crude sewage and the effluents from the 4-foot bed, it is to be noted that the number of spores of *B. enteritidis* varied from 10 to 1,000 per c.c. In the case of both the 6-foot coke-bed effluent and the effluent from the laboratory vessel the numbers were found to vary from 10 to 100 per c.c., but there may have been more spores present, as the minimum amount of the liquid added to the milk-tubes was 0·01 c.c.

In comparing the various results, it is noteworthy that although the number of spores of *B. enteritidis* in the crude sewage exceeded those found in the 4-foot coke-bed effluent on two separate occasions, the number on three other occasions was greatest in the 4-foot coke-bed effluent. In the remaining five comparative experiments the numbers were approximately equal. As regards the crude sewage and the 6-foot coke-bed effluent, the numbers were equal on July 6th, 20th, and 27th. On August 4th the number was greater in the crude sewage, and on August 9th greatest in the 6-foot coke-bed effluent. The laboratory coke-bed effluent and the crude sewage gave an equal number on July 20th and August 9th, but on July 27th and August 4th the crude sewage contained a larger number of spores.

Judging these results as a whole, it cannot be said that the biological processes at work in the coke-beds produced any significant alteration in the number of spores of this pathogenic anaërobe. This is the less to be regretted since the effluents are discharged into a large tidal river* below looks, the water of which is not used for drinking purposes. Still, it is to be thought of that the cultures of *B. enteritidis sporogenes* are extremely virulent, and that Dr. Klein's results seem to prove that this anaërobe may be causally related to acute diarrhœa. At all events, it is highly important from a practical as well as from a scientific point of view to continue these observations on the number of spores of *B. enteritidis* in crude sewage and in the effluents from the coke-beds.

The method for detecting the presence of the spores of this bacillus is as follows†:—Dilute 1 part of crude sewage or of effluent, as the case may be, with 99 parts of sterile water; of this dilution add 1·0, 0·1, and 0·01 c.c. severally to three sterile milk tubes. Heat the tubes to 80° C. for ten minutes, and cultivate anaërobically by Buchner's method at a temperature of 37° C. In certain cases it is necessary to add as much as 0·1 c.c. of the crude sewage or effluent directly to the milk tube. When *B. enteritidis* is present the casein is precipitated, the whey remains nearly colourless, and there is a marked development of gas.‡ These changes in the milk commonly take place in less than 24 hours. A guinea-pig inoculated subcutaneously with 1 c.c. of the whey, usually dies in less than 24 hours, and presents, on post-mortem examination, appearances which are typical of enteritidis (extensive gangrene, sanguineous exudation full of bacilli, etc.).

(a) *Bacillus Coli Communis*.§

In the First Report the results of a considerable number of experiments were given, showing that *B. coli* was present in the crude sewage in numbers exceeding 100,000 per c.c., and it was stated that in future

* That Thames water itself contains the spores of *B. enteritidis sporogenes* may be seen by referring to Table II.
† See page 3 of First Report.   ‡ See No. 14, plate I. of First Report.
§ In the First Report a number of illustrations were shown, illustrating the morphological and biological characters of this micro-organism.

work this micro-organism would be searched for in the effluents from the coke-beds. It was also pointed out that its presence in such an effluent might be considered as showing a derivation of the liquid from sewage, and hence as indicating the possible presence of other and, perhaps, dangerous bacteria which had survived the biological processes at work in the coke-beds, but did not signify that the liquid was necessarily of a degraded, offensive, and putrescible character.

Although it has been long known that *B. coli* is an organism characteristic of fæcal discharges, no accurate record exists of the determination of the number of these germs in samples of crude sewage periodically examined. The importance of establishing such a record will be understood when it is remembered that sewage is the chief and most dangerous source of pollution of potable waters. If, then, we know the number of germs of *B. coli* in crude sewage, and if we find that pure waters do not contain this organism, or contain it only in few numbers, we are in a position to judge of the importance of its presence in a given sample of water. It is true that *B. coli* is widely distributed in nature, is capable under certain circumstances of multiplying outside the animal body, and is present in other excreta besides those of human beings. These facts no doubt to some extent lessen the value of the mere demonstration of *B. coli* in water, but sink into comparative insignificance when the relative abundance of this micro-organism is considered. It has already been shown that *B. coli* may be present in crude sewage in numbers exceeding 100,000 per c.c., and it may safely be affirmed that in a water free from any likelihood of pollution *B. coli* is not discoverable, or is present in very few numbers. A minimal quantity of sewage gaining access to a large bulk of water would increase the number of *B. coli* in water very greatly; and it is difficult to conceive of any substance other than sewage producing a result in any way comparable. It is particularly difficult to conceive of a substance of an unobjectionable nature producing this result. No doubt in certain cases the presence of *B. coli* in water may arise from *indirect* pollution with sewage, as, for example, where the surface drainage from manured or polluted soil gains access to a supply. Yet, even here, the danger so far as potability is concerned could hardly be considered remote. It is not the presence of *B. coli* in water, but its relative abundance, which entitles the bacteriologist to assume pollution direct or indirect with sewage, or to avoid all controversial points with a substance of an objectionable nature. Certainly, in view of the fact that this organism is not discoverable in waters free from any likelihood of sewage pollution, and is found to be present in great numbers in waters which are known to be fæcally contaminated, and is present in crude sewage in numbers vastly exceeding any other substance at all likely to contaminate a water supply; it seems justifiable to conclude that its presence in considerable numbers in a given sample of water indicates pollution of an objectionable kind, and, in all probability, points to contamination with sewage.

Enough, at all events, has been made out to show the importance of a large number of observations having for their object the collection of data dealing not only with the presence, but with the relative abundance of *B. coli* in sewage. Of the importance of experiments having for their aim the estimation of the number of *B. coli* in the effluents from the biological coke-beds as compared with the number in the crude sewage applied to the coke-beds there cannot be a doubt. And here it may be stated that much that has been said in the above sentences applies with perhaps even greater force to *B. enteritidis sporogenes*. *B. coli* is an example of an aërobic, and *B. enteritidis sporogenes* of an anaërobic micro-organism, the former being peculiarly abundant in crude sewage, and the latter less abundant, but perhaps more characteristic. The estimation of their numbers in the effluents as compared with the crude sewage needs no comment to show its importance.

The following is a summary of the results, as regards *B. coli* shown in col. 5, Table I.—

Date.	Crossness crude sewage.	Effluent from 4-foot coke-bed.	Effluent from 6-foot coke-bed.
1898. 11th May	No colonies of B. coli in phenol gelatine plate containing 0·00001 c.c. sewage (expt. 1)	200,000 B. coli per c.c. (expt. 2)	
18th May	Gas forming coli bacteria, 300,000 per c.c. (expt. 3)	200,000 gas forming coli per c.c. (expt. 4)	
25th May	Gas forming coli, 1,500,000 per c.c. (expt. 5)	Gas forming coli, 700,000 per c.c. (expt. 6)	
9th June	Gas forming coli, 200,000 per c.c. (expt. 7)	Gas forming coli, 100,000 per c.c. (expt. 8)	
15th June	Gas forming coli, 300,000 per c.c. (expt. 9)	Gas forming coli, 600,000 per c.c. (expt. 10)	
22nd June	Gas forming coli, 300,000 per c.c. (expt. 11)	Gas forming coli, 300,000 per c.c. (expt. 12)	
20th July	Gas forming coli, at least 300,000 per c.c. (expt. 15)	Gas forming coli, at least 300,000 per c.c. (expt. 16)	
27th July	B. coli, 500,000 per c.c. (expt. 19)	...	B. coli, 600,000 per c.c. (expt. 20)
4th August	B. coli, 200,000 per c.c. (expt. 23)	B. coli, 300,000 per c.c. (expt. 24)	
9th August	B. coli, 1,000,000 per c.c. (expt. 27)	...	B. coli, 200,000 per c.c. (expt. 28)

These results are shown in graphic form in Diagram 9.

It is to be noted that on three separate occasions the number of *B. coli* in the crude sewage exceeded the number found in the 4-foot coke-bed effluent, and on three other occasions the conditions were reversed in this respect. In the two comparative experiments remaining the numbers were approximately equal.

As regards the 6-foot coke-bed effluent, on one occasion the number in the effluent was somewhat greater than the number in the crude sewage; on the other occasion the crude sewage contained a much larger number.

Judging the experiments as a whole it cannot be said that the biological processes at work in the coke-beds effected any marked alteration in the number of *B. coli*. It must not, however, be too lightly considered that this implies that the effluent was necessarily of an offensive and putrescible character. *B. coli* and other putrefactive bacteria no doubt work in the direction of purifying the sewage, and their presence in the effluent might only mean that the purification had not been carried sufficiently far to allow of a decrease in their numbers, owing to the incomplete reduction of the organic matters on which they feed and which allow of their continued multiplication. Yet, when this has been said, it must also be admitted that the passage of an aërobic non-spore-forming bacillus typical of excremental matters through the coke-beds, in practically unaltered numbers, is not a desirable state of things. It is true that *B. coli* is not pathogenic in the ordinary meaning of the word, but its presence in the effluents implies the possible presence of other bacteria—it might be of dangerous sort. Still, on the whole it may be said that the balance of evidence points to pathogenic aërobic bacteria being liable to be crowded out in the struggle for existence in a nutrient fluid containing a mixed bacterial flora and one rich in saprophytic micro-organisms.*
Lastly, it must be remembered that the effluent is discharged into a large tidal river at a point far below the lowest "intake" of water for waterworks purposes. Moreover, the Thames before it reaches the Outfalls of the Sewage Works is already grossly polluted with excremental matters.†

In searching for *B. coli* in the crude sewage and in the effluents the following plan was adopted‡:—10 c.c. of sterile gelatine, contained in a test tube, were melted, 0·1 c.c. of five per cent. phenol added, and then the gelatine was poured into a Petri's capsule and allowed to become quite solid. 0·1 c.c. of diluted sewage, or else of effluent (1 : 10,000) was next added and spread over the entire surface of the gelatine with a platinum spreader. Colonies which were typical of *B. coli* in their microscopical appearance and in the manner of their growth were then subcultured in broth (for diffuse cloudiness and indol reaction), in litmus milk (for acidity and clotting), and in gelatine shake culture (for gas formation). It was not, however, found possible in all of the experiments to apply all of these tests, although, in the majority of cases the gas test in gelatine was applied.

"(*c*) OTHER SPECIES OF BACTERIA.

Besides searching for *B. enteritidis* and *B. coli* the attempt was also made to estimate the number as well as the character of other microbes present in the crude sewage and in the effluents. Notes under this heading will be found in col. 5 of Table I. Thus the organism called "*sewage proteus*," of which a description appears in this report, was found to be present in great numbers (usually over 100,000 per c.c.) in both the crude sewage and in the effluents. Other microbes found in the crude sewage and effluents were *B. fluorescens liquefaciens* and its varieties, *B. fluorescens non-liquefaciens*, *B. mesentericus* (a description of two varieties of this bacillus is given in this report), *B. subtilis*, *B. mycoides*, *B. pyocyaneus*,§ streptococci, staphylococci, &c.

In experiments 3 and 5 (col. 5, Table I.) the number of bacteria in the crude sewage capable of growing at 37° C. in agar was estimated. The numbers were 1,260,000 and 1,171,000 per c.c. as compared with 3,670,000 and 6,400,000 obtained by gelatine plate cultivation at 20° C. A similar experiment (experiment 4) with the effluent from the 4-foot filter bed gave 1,630,000 (agar, at 37° C.) as compared with 4,100,000 (gelatine, at 20° C.) bacteria per c.c.

In experiments 9 and 10 (col. 5, Table I.) it was sought to discover the smallest amount of crude sewage and of effluent which in broth cultures at 20° C. would produce growth, indol reaction and offensive smell. No growth occurred in either case, when as little as 0·0000001 c.c. was inoculated into the broth; but when 0·000001 c.c. was used, growth occurred both in the case of the crude sewage and of the effluent, and the cultures had an offensive smell and gave indol reaction.

In concluding this section of the report it may be said that when making comparative cultivations from different liquids the trained observer can often detect differences in the characters of the colonies developing in the nutrient media which are none the less real because they cannot always be put into definite language. Speaking from this point of view, it must be admitted that little or no real distinction could be made out between the cultures made from the crude sewage and those made from the effluents, other than those points of difference already considered—namely, a slight reduction in the total number of aërobic bacteria, the number of spores of aërobic bacteria, and the number of liquefying aërobic bacteria found in the effluent, as compared with the crude sewage.

---

* See, however, notes under heading—The biological treatment of sewage.
† See results shown in Table II. ‡ See pages 3 and 6 of First Report.
§ A cultivation of *B. pyocyaneus* isolated from a sample of Crossness crude sewage proved to be extremely virulent. Thus 1 c.c. of a 24 hours' broth culture (at 37° C.) injected subcutaneously into a guinea-pig, killed the animal in less than 24 hours, and the organism was recovered in pure culture from the heart's blood, spleen, &c.

IV.—TABLES AND DIAGRAMS DEALING WITH THE RESULTS OF THE BACTERIOLOGICAL EXAMINATION OF THE CRUDE SEWAGE, OF THE EFFLUENTS FROM THE COKE-BEDS, OF THE EFFLUENTS FROM THE CHEMICAL PRECIPITATION WORKS, AND OF SAMPLES OF THAMES WATER.

TABLE I., showing the results of the bacteriological examination of Crossness crude sewage, the effluent from the 4-foot coke-bed, the effluent from the 6-foot coke-bed, and the effluent from the laboratory vessel (effluent from 6-foot coke-bed again treated in the laboratory at Crossness).

Expt.	Description of the samples.	Total number of bacteria in 1 c.c. of the sample.	Number of spores of bacteria in 1 c.c. of the sample.	Number of bacteria causing liquefaction of the gelatine in 1 c.c. of the sample.	Remarks.
1.	2.	3.	4.	5.	
1	Crossness crude sewage, May 11, 1898.	3,930,000 (average of two expts. with 0·0001 and 0·00001 c.c. sewage).	460 (one expt. with 0·1 c.c. sewage).	400,000 (one expt. with 0·00001 c.c. sewage).	Spores of B. enteritidis present in milk culture containing 0·1 c.c. sewage; absent in cultures containing 0·01, 0·001, and 0·0001 c.c. sewage. In phenol gelatine plate containing 0·00001 c.c. sewage, no colonies of B. coli. B. fluorescens liquefaciens, 200,000 in 1 c.c. "Sewage proteus," 200,000 in 1 c.c. In cultivation for spores (0·1 c.c. sewage, col. 3) 10 colonies of B. mesentericus and several of B. subtilis. At least 5 different species of bacteria in 0·00001 c.c. sewage.
2	Crossness effluent from 4-foot coke-bed, May 11, 1898.	4,800,000 (average of two expts. with 0·0001 and 0·00001 c.c. effluent).	260 (one expt. with 0·1 c.c. effluent).	1,300,000 (one expt. with 0·00001 c.c. effluent).	Spores of B. enteritidis present in 0·01 c.c.; absent in 0·001 and 0·0001 c.c. effluent. In phenol gelatine plate containing 0·00001 c.c. effluent, at least two colonies of B. coli. No fluorescent colonies noted in any of the cultures on the fifth day. "Sewage proteus," 400,000 in 1 c.c. In cultivation for spores (0·1 c.c. effluent, col. 3) 20 colonies of B. mesentericus. At least four different species of bacteria in 0·00001 c.c. effluent.
3	Crossness crude sewage, May 18, 1898.	3,670,000 (average of two expts. with 0·0001 and 0·00001 c.c. sewage).	200 one expt. with 0·1 c.c. sewage).	100,000 (one expt. with 0·00001 c.c. sewage).	Spores of B. enteritidis present in 0·1, 0·01, and 0·001 c.c. sewage. An agar plate culture incubated at 37° C. yielded only 1,260,000 bacteria per c.c. sewage as compared with 3,670,000 by gelatine plate culture at 20° C. Gas-forming coli bacteria, 200,000 in 1 c.c. sewage. B. fluorescens liquefaciens, 100,000 in 1 c.c. sewage. "Sewage proteus," 100,000 in 1 c.c. sewage.
4	Crossness effluent from 4-foot coke-bed, May 18, 1898.	4,100,000 (average of two expts. with 0·0001 and 0·00001 c.c. effluent).	140 (one expt. with 0·1 c.c. effluent).	700,000 (one expt. with 0·00001 c.c. effluent).	Spores of B. enteritidis present in 0·1 and 0·01 c.c.; absent in 0·001 c.c. effluent. An agar plate culture incubated at 37° C. yielded only 1,630,000 bacteria per c.c. effluent as compared with 4,100,000 by gelatine plate culture at 20° C. Gas-forming coli, at least 200,000 per c.c. effluent. B. fluorescens liquefaciens 10,000, and B. fluorescens non-liquefaciens 10,000 in 1 c.c. effluent. "Sewage proteus," 100,000 in 1 c.c. effluent.
5	Crossness crude sewage, May 25, 1898.	6,400,000 (one expt. with 0·00001 c.c. sewage).	370 (one expt. with 0·1 c.c. sewage).	900,000 (one expt. with 0·00001 c.c. sewage).	The liquefaction of the gelatine with 0·0001 c.c. sewage (col. 2) was so great that this culture could not be counted; the cultivation, however, containing 0·00001 c.c. sewage was counted. Spores of B. enteritidis in 0·1 and 0·01 c.c.; absent in 0·001 c.c. sewage. Gas-forming coli, 1,500,000 in 1 c.c. sewage (see photo 1, Plate I., First Report). An agar plate culture incubated at 37° C. yielded 1,171,000 colonies per c.c. sewage. "Sewage proteus," at least 100,000 per c.c. sewage. No fluorescent colonies noted in culture containing 0·00001 c.c. sewage. In cultivation for spores (0·1 c.c. sewage, col. 3) several colonies of B. mesentericus

Expt.	Description of the samples. 1.	Total number of bacteria in 1 c.c. of the sample. 2.	Number of spores of bacteria in 1 c.c. of the sample. 3.	Number of bacteria causing liquefaction of the gelatine in 1 c.c. of the sample. 4.	Remarks. 5.
6	Crossness effluent from 4-foot coke-bed, May 25, 1898.	6,100,000 (one expt. with 0·00001 c.c. effluent.)	380 (one expt. with 0·1 c.c. effluent).	700,000 (one expt. with 0·00001 c.c. effluent).	The liquefaction of the gelatine with 0·00001 c.c. effluent was so rapid that the culture could not be counted; the cultivation, however, containing 0·000001 c.c. effluent was counted. Spores of *B. enteritidis* present in 0·1, 0·01 and 0·001 c.c. effluent. Dr. Klein inoculated a guinea-pig with 1 c.c. of the whey from an anaërobic milk culture containing 0·0001 c.c. effluent. The following day the animal was dead, and presented on examination the appearances which are typical of enteritidis (extensive gangrene, sanguineous exudation full of bacilli, etc.). Gas-forming coli, 700,000 per c.c. of effluent. No fluorescent colonies noted in 0·00001 c.c. effluent (5th day).
7	Crossness crude sewage, June 9, 1898.	5,500,000 (one expt. with 0·00001 c.c. sewage).	550 (one expt. with 0·1 c.c. sewage).	1,400,000 (one expt. with 0·00001 c.c. sewage).	Spores of *B. enteritidis* present in 0·1 and 0·01 c.c. sewage. In the cultivation for spores (0·1 c.c. sewage, col. 3), several colonies of *B. mesentericus*. Gas-forming coli, 200,000 in 1 c.c. sewage. No fluorescent colonies noted in any of the cultures (4th day). "Sewage proteus," at least 100,000 per c.c. of sewage.
8	Crossness effluent from 4-foot coke-bed, June 9, 1898.	1,200,000 (average of three expts. with 0·00001 effluent in each case).	230 (one expt. with 0·1 c.c. effluent).	200,000 (one expt. with 0·00001 c.c. effluent).	Spores of *B. enteritidis* present in 0·1 and 0·01 c.c. effluent. Gas-forming col., 100,000 in 1 c.c. effluent. No fluorescent colonies noted in any of the cultures (4th day). "Sewage proteus," 100,000 in 1 c.c. effluent. Anaërobic grape sugar gelatine cultivations were made with similar amounts of effluent (expt. 8) and crude sewage (expt. 7); the colonies were certainly more numerous in the culture made from the crude sewage.
9	Crossness crude sewage, June 15, 1898.	4,000,000 (one expt. with 0·00001 c.c. sewage).	180 (one expt. with 0·1 c.c. sewage).	600,000 (one expt. with 0·00001 c.c. sewage).	Spores of *B. enteritidis* present in 0·1 c.c. sewage. In the cultivation for spores (0·1 c.c. sewage, col. 3), 1 colony of *B. mycoides*. Gas-forming coli, 300,000 per c.c. sewage. *B. fluorescens liquefaciens*, 100,000 in 1 c.c. sewage. (a) 0·0001; (b) 0·000001; and (c) 0·0000001 c.c. sewage inoculated into each of three bouillon tubes and incubated at 20° C.; on 7th day no growth in (c), growth and offensive smell in (a) and (b)—(a) gave strong indol reaction and (b) faint trace.
10	Crossness effluent from 4-foot coke-bed, June 15, 1898.	5,300,000 (one expt. with 0·00001 c.c. effluent).	300 (one expt. with 0·1 c.c. effluent).	1,100,000 (one expt. with 0·00001 c.c. effluent).	Spores of *B. enteritidis* present in 0·1 c.c. effluent. Gas-forming coli, 600,000 per c.c. effluent. (a) 0·0001; (b) 0·000001; and (c) 0·0000001 c.c. effluent inoculated into each of three bouillon tubes and incubated at 20° C.; on 7th day no growth in (c), growth and offensive smell in (a) and (b)—(a) and (b) both gave distinct indol reaction.
11	Crossness crude sewage, June 22, 1898.	9,100,000 (one expt. with 0·00001 c.c. sewage).	310 (one expt. with 0·1 c.c. sewage).	900,000 (one expt. with 0·00001 c.c. sewage).	Spores of *B. enteritidis* present in 0·1, 0·01 and 0·001 c.c. sewage. Gas-forming coli, 300,000 in 1 c.c. sewage. "Sewage proteus," 100,000 per c.c. sewage.
12	Crossness effluent from 4-foot coke-bed, June 22, 1898.	3,000,000 (one expt. with 0·00001 c.c. effluent.)	60 (one expt. with 0·1 c.c. effluent).	600,000 (one expt. 0·00001 c.c. effluent).	Spores of *B. enteritidis* present in 0·1 c.c.; absent in 0·01 and 0·001 c.c. effluent. No colonies of "sewage proteus" in 0·00001 c.c. effluent. Gas-forming coli, 300,000 per c.c. effluent. *B. fluorescens liquefaciens*, 100,000 in 1 c.c. effluent.

Expt.	Description of the samples.	Total number of bacteria in 1 c.c. of the sample.	Number of spores of bacteria in 1 c.c. of the sample.	Number of bacteria causing liquefaction of the gelatine in 1 c.c. of the sample.	Remarks.
1.		2.	3.	4.	5.
13	Crossness crude sewage, July 6, 1898.	This sample was examined for B. enteritidis only.			Spores of B. enteritidis present in 0·1 c.c.; absent 0·01 and 0·001 c.c. sewage.
14	Crossness effluent from 6-foot coke-bed, July 6, 1898.	do.			Spores of B. enteritidis present in 0·1 c.c.; absent in 0·01 and 0·001 c.c. sewage.
15	Crossness crude sewage, July 20, 1898.	12,800,000 (one expt. with 0·00001 c.c. sewage).	400 (one expt. with 0·1 c.c. sewage).	1,700,000 (one expt. with 0·00001 c.c. sewage).	The liquefaction of the gelatine with 0·0001 c.c. sewage (col. 2) was so rapid that the colonies could not be counted. The cultivation containing 0·00001 c.c. sewage was, however, counted. Spores of B. enteritidis present in 0·1 and 0·01 c.c.; absent in 0·001 c.c. sewage. Gas-forming coli, at least 300,000 per c.c. of sewage. B. fluorescens liquefaciens, 100,000 in 1 c.c. sewage. "Sewage proteus," several colonies in 0·00001 c.c. sewage. Spores of B. mesentericus and B. subtilis present in 0·1 c.c. sewage.
16	Crossness effluent from 4-foot coke-bed, July 20, 1898.	9,200,000 (one expt. with 0·00001 c.c. effluent).	430 (one expt. with 0·1 c.c. effluent).	1,000,000 (one expt. with 0·00001 c.c. effluent).	The liquefaction of the gelatine with 0·0001 c.c. effluent (col. 2) was so rapid that the colonies could not be counted; the cultivation, however, containing 0·00001 c.c. sewage was counted. Spores of B. enteritidis present in 0·1 and 0·01; absent in 0·001 c.c. effluent. Gas-forming coli at least 300,000 per c.c. of effluent. B. fluorescens liquefaciens, 100,000 in 1 c.c. effluent. Spores of B. mesentericus and B. subtilis present in 0·1 c.c. effluent.
17	Crossness effluent from 6-foot coke-bed, July 20, 1898.	This sample was examined for B. enteritidis only.			Spores of B. enteritidis present in 0·1 and 0·01 c.c. effluent.
18	Crossness effluent from laboratory vessel (6-foot coke-bed effluent again treated in laboratory), July 20, 1898.	do.			Spores of B. enteritidis present in 0·1 and 0·01 c.c. effluent.
19	Crossness crude sewage, July 27, 1898.	7,200,000 (one expt. with 0·00001 c.c. sewage).	870 (one expt. with 0·1 c.c. sewage).	900,000 (one expt. with 0·00001 c.c. sewage).	Spore of B. enteritidis in 0·1 and 0·01 c.c. sewage. B. coli, 500,000 in 1 c.c. sewage. Several colonies of "sewage proteus" in 0·00001 c.c. sewage. B. fluorescens liquefaciens 100,000 per c.c. sewage. One colony of B. mycoides in plate culture for spores (0·1 c.c. sewage, col. 3).
20	Crossness effluent from 6-foot coke-bed, July 27, 1898.	6,600,000 (one expt. with 0·00001 c.c. effluent).	480 (one expt. with 0·1 c.c. effluent).	300,000 (one expt. with 0·00001 c.c. effluent).	Spores of B. enteritidis in 0·1 and 0·01 c.c. effluent. B. coli, 600,000 in 1 c.c. effluent.
21	Crossness effluent from 4-foot coke-bed, July 27, 1898.	This sample was examined for B. enteritidis only.			Spores of B. enteritidis present in 0·1 and 0·01 c.c.; absent in 0·001 c.c. effluent.

Expt.	Description of the samples. 1.	Total number of bacteria in 1 c.c. of the sample. 2.	Number of spores of bacteria in 1 c.c. of the sample. 3.	Number of bacteria causing liquefaction of the gelatine in 1 c.c. of the sample. 4.	Remarks. 5.
22	Crossness effluent from laboratory vessel (6-foot coke-bed effluent again treated in laboratory), July 27, 1898.	This sample was examined for *B. enteritidis* only.			Spores of *B. enteritidis* present in 1·0 and 0·1 c.c.; absent in 0·01 c.c. effluent.
23	Crossness crude sewage, Aug. 4, 1898.	4,200,000 (one expt. with 0·00001 c.c. sewage).	280 (one expt. with 0·1 c.c. sewage).	400,000 (one expt. with 0·00001 c.c. sewage).	Spores of *B. enteritidis* in 0·1 and 0·01 c.c. crude sewage. *B. coli*, 200,000 in 1 c.c. crude sewage. "*Sewage proteus*" 200,000 in 1 c.c. crude sewage.
24	Crossness effluent from 4-foot coke-bed, Aug. 4, 1898.	1,800,000 (one expt. with 0·00001 c.c. effluent).	220 (one expt. with 0·1 c.c. effluent).	500,000 (one expt. with 0·00001 c.c. effluent).	Spores of *B. enteritidis* in 0·1 and 0·01 c.c. effluent. *B. coli*, 300,000 in 1 c.c. effluent. "*Sewage proteus*," 200,000 in 1 c.c. effluent.
25	Crossness effluent from 6-foot coke-bed, Aug. 4, 1898.	This sample was examined for *B. enteritidis* only.			Spores of *B. enteritidis* in 0·1 c.c. effluent.
26	Crossness effluent from laboratory vessel (6-foot coke-bed effluent again treated in laboratory), Aug. 4, 1898.		do.		Spores of *B. enteritidis* in 0·1 c.c. effluent.
27	Crossness crude sewage, Aug. 9, 1898.	3,600,000 (one expt. with 0·00001 c.c. sewage).	340 (one expt. 0·1 c.c. sewage).	1,100,000 (one expt. with 0·00001 c.c. sewage).	Spores of *B. enteritidis* in 0·1 c.c.; not in 0·01 c.c. sewage. *B. coli*, 1,000,000 in 1 c.c. sewage.
28	Crossness effluent from 6-foot coke-bed, Aug. 9, 1898.	1,700,000 (one expt. with 0·00001 c.c. effluent).	300 (one expt. with 0·1 c.c. effluent).	200,000 (one expt. with 0·00001 c.c. effluent).	Spores of *B. enteritidis* present in 0·1 and 0·01 c.c. effluent. *B. coli*, 200,000 in 1 c.c. sewage.
29	Crossness effluent from 4-foot coke-bed, Aug. 9, 1898.	This sample was examined for *B. enteritidis* only.			Spores of *B. enteritidis* in 0·1 and 0·01 c.c. effluent.
30	Crossness effluent from laboratory vessel (6-foot coke-bed effluent again treated in laboratory), Aug. 9, 1898.		do.		Spores of *B. enteritidis* in 0·1; not in 0·01 c.c. effluent.

TABLE 2.—Showing the results of the bacteriological examination of the effluents from the Crossness and Barking Outfall Works and the water of the River Thames.

Expt.	Description of the samples. 1.	Total number of bacteria in 1 c.c. 2.	Number of spores of bacteria in 1 c.c. 3.	Number of bacteria causing liquefaction of the gelatine in 1 c.c. 4.	Remarks. 5.
1	Effluent from chemical precipitation works at Barking, Nov. 2, 1898.	5,200,000 (two expts. with 0·0001 and 0·00001 c.c. effluent).	100 (one expt. with 0·1 c.c. effluent).	1,100,000 (one expt. with 0·0001 c.c. effluent).	Spores of *B. enteritidis* present in 0·1, 0·01 c.c. absent in 0·001 c.c. Five colonies indistinguishable from *B. coli* in phenol gelatine plate containing 0·00001 c.c. effluent (500,000 per c.c. of effluent). The plate, containing 0·0001 c.c. effluent (col. 2), was too crowded to allow of the colonies being accurately counted. The plate, containing 0·1 c.c. effluent (col. 3), contained some rapidly liquefying species, so that the numbers given are only approximate.
2	Effluent from chemical precipitation works at Crossness, Nov. 2, 1898.	9,600,000 (two expts. with 0·0001 and 0·00001 c.c. effluent).	300 (one expt. with 0·1 c.c. effluent).	1,000,000 (one expt. with 0·0001 c.c. effluent).	Spores of *B. enteritidis* present in 0·1 and 0·01 c.c.; absent in 0·001 c.c. Twenty-six colonies indistinguishable from *B. coli* in phenol gelatine plate containing 0·00001 c.c. effluent (2,600,000 per c.c. of effluent). The 0·00001 c.c. plate (col. 2) was counted, but not the 0·0001 c.c. plate, as this was too crowded.
3	Thames water collected off Greenhithe at about 2 p.m., October 12th, 1898, on a tide which had ebbed from Crossness for about two hours. The sample was, therefore, judged to be free from admixture with recent sewage effluent. (Dry weather.)	10,000	63	No record.	At least 1 but less than 10 spores of *B. enteritidis* per c.c. of water. No colonies of *B. coli* observed in phenol gelatine plate containing as small an amount of water as 0·001 c.c. That *B. coli* was present, however, in the water need not be doubted. There were present in the cultures numerous colonies of *B. fluorescens liquefaciens* and non-liquefaciens, *B. mesentericus*, *B. mycoides*, and proteus forms were also found.
4	Thames water, taken at low tide at Barking, and outside the influence of sewage discharge, as far as this was possible, Nov. 3, 1898. (Dry weather.)	34,600	89	(A Surface gelatine plate culture containing 0·1 c.c. of the water was completely liquefied by the 2nd day).	At least 1 but less than 10 spores of *B. enteritidis* per c.c. of water. Eighty-four colonies, indistinguishable from *B. coli*, in phenol gelatine plate containing 0·1 c.c. of the water. There were present in the cultures, colonies of *B. mesentericus*, proteus senkeri, *B. fluorescens liquefaciens* etc. Dr. Klein inoculated a guinea-pig with 1 c.c. of milk culture (containing 1·0 c.c. of water) which showed the typical enteritidis change. The animal died, and presented upon post-mortem examination the usual appearances; sanguineous exudation full of bacilli, etc.
5	Thames water, sample taken in mid-stream between Sunbury and Hampton, just above intake of the Southwark and Vauxhall Water Co. 10.40 a.m., Nov. 15, 1898 (dry weather).	5,100	56	No record.	Spores of *B. enteritidis* sporogenes absent in 1·0, 5·0, 10·0, 100·0 c.c.; present in 300 c.c. Dr. Klein inoculated a guinea-pig with 1 c.c. from latter culture. The animal died in less than 24 hours, and on post-mortem examination presented the usual appearances—swollen belly, sanguineous exudation swarming with bacilli, &c. Forty gas-forming *B. coli* per c.c. of water. Colonies of *B. fluorescens liquefaciens* and non-liquefaciens and proteus-forms present in the cultures. Also a large number of colonies of *B. aquatilis sulcatus* type. The presence of *B. enteritidis* in 300 c.c. of the water was demonstrated by a special filtration process.

Expt.	Description of the samples. 1.	Total number of bacteria in 1 c.c. 2.	Number of spores of bacteria in 1 c.c. 3.	Number of bacteria causing liquefaction of the gelatin in 1 c.c. 4.	Remarks. 5.
6	Thames water, sample taken in mid-stream at Twickenham, opposite Ham house, i.e., about half-way between Glover's island and Eel Pie island. 11 a.m., Nov. 29, 1898 (very wet weather).	3,000	18	No record.	Spores of *B. enteritidis sporogenes* present in 10 c.c., 5 c.c., 1 c.c., 0·1 c.c. of the water; absent in 0·01 c.c. In a phenol gelatine plate culture containing 0·01 c.c. of the water, one colony indistinguishable from *B. coli*. Subcultures from this colony gave the following results: Gas in 24 hours in gelatine shake culture incubated at 20° C. Diffuse cloudiness in broth in 24 hours at 37° C. No indol in five days. Slight acidity but no clotting in litmus milk in 48 hours at 37° C. In 96 hours, however, the milk showed a solid clot. In an ordinary gelatine plate culture containing 0·01 c.c. of the water, one colony indistinguishable from *B. coli*. Subcultures from this colony gave the following results: Gas in 24 hours at 20° C. in gelatine shake culture, diffuse cloudiness in broth in 24 hours at 37° C.; acidity but no clotting in litmus milk culture in 24 hours at 37° C. No indol in broth culture, fourth day, at 37° C.

V.—DESCRIPTION OF SOME OF THE BACTERIA FOUND IN THE CRUDE SEWAGE AND IN THE EFFLUENTS FROM THE COKE-BEDS.

1. B. Coli Communis.
2. B. Mesentericus.
   Sewage Variety E.
   Sewage Variety I.
3. Sewage Proteus.
4. B. Frondosus.
5. B. Fusiformis.
6. B. Subtilissimus.
7. B. Subtilis.
   Sewage Variety A.
   Sewage Variety B.
8. B. Membraneus Patulus.
9. B. Capillareus.

1. Bacillus Coli Communis.

[*An aërobic, non-chromogenic, slightly motile, non-liquefying bacillus.*]

Source.—The excreta of human beings and many of the lower animals. Very abundant in London crude sewage; usually more than 100,000 per c.c. of sewage.

Morphology.†—Small bacilli with rounded ends, hardly longer than broad; solitary, often in pairs; rarely in chains containing more than two segments.

Motility.—Usually only feebly motile.

Spore formation.—No spores are formed.

Flagella.—Dr. M. H. Gordon gives the flagella average as 1 to 3.

Temperature.—Grows best at 37° C., but also very well at the ordinary temperature.

Gelatine plate cultures.¶—The colonies develop in from 24 to 48 hours at 20° C. The deep colonies are not characteristic, the surface colonies peculiarly so, appearing as delicate, slightly granular films, of an irregularly circular shape, which are bluish-white by reflected and of an amber colour by transmitted light; they are transparent, and sometimes iridescent, especially towards the periphery, but at the centre and over the entire surface in old cultures an opacity due to a greater thickness of the bacterial growth is observed; later these surface colonies may become marked by concentric, or radiating, or irregular markings. The surrounding gelatine frequently acquires a dull, cloudy, faded appearance. The gelatine is not liquefied.

Gelatine "slab" cultures.—The growth on the surface is like a surface colony in a plate culture, but tends to be more luxuriant, due to the greater thickness of the medium. A white growth extends to the foot of the stab, and gas fissures frequently appear in the gelatine. The gelatine, as already stated, is not liquefied.

Gelatine "streak" cultures.—The growth is like an elongated surface colony in a gelatine plate culture, but is, perhaps, more luxuriant. Briefly, a delicate faintly-granular film forms, with transparent and irregular margins. Down the centre, longitudinally, the growth is more opaque. Sometimes the film shows iridescence, and in old cultures it may become irregularly thickened. The gelatine, which is not liquefied, becomes often clouded. The growth, which is bluish-white by reflected light, has a yellowish-amber colour by transmitted light.

Potato-gelatine plate, "slab" and "streak" cultures.—The growth is somewhat similar in appearance to the above, but tends to be more circumscribed, is slower, and of a characteristic brown colour.

Phenol (0·05 per cent.) *gelatine cultures.**—These do not differ from ordinary gelatine cultures except that growth is delayed.

Gelatine "shake" cultures.—Numerous gas bubbles are formed, usually in 24 hours, at 20° C.

Agar plate cultures.—The growth is not so characteristic as in gelatine. The superficial colonies have a moist glistening white appearance.

† ¶ * See figs. 6, 4, 13 and 5, Plate I. of First Report.

Agar "stab" cultures.—Growth occurs all the way down the stab, and on the surface a white layer is developed.

Agar "streak" cultures.—At 37° C. the growth is very rapid, and occurs as an abundant moist and white layer.

Potato cultures.—A rich yellowish-brown layer quickly develops.

Broth cultures.—The growth is characteristic. In less than 24 hours, at 37° C., the broth is uniformly turbid. Later, a heavy bacterial deposit collects at the foot of the tube. There is no distinct pellicle formation, but sometimes an imperfect scum forms on the surface.

Phenol (0·05 per cent.) broth cultures.—The growth is the same as in an ordinary broth cultivation.

Litmus milk cultures.—The growth is extremely characteristic. Usually, an acid solid clotting of the milk takes place in 24 hours at 37° C. Occasionally the clotting is somewhat delayed. First of all, the bluish-purple colour changes to pink; then clotting occurs, and the milk, except at the free surface, becomes white. Later, the redness extends from the surface downwards until the whole contents of the tube are bright red in colour.

Blood serum cultures.—An abundant white layer is quickly developed at 37° C.; there is no liquefaction.

Indol reaction.—Indol reaction is usually well marked in broth cultures kept at 37° C. for five days.

Reduction of nitrates.—In 24 hours, at 37° C., reduction of nitrates to nitrites well marked. [Broth 5 per cent. KNO_3 0·1 per cent., water 94·9 per cent.]

Widal's test.—Typhoid blood serum gives a negative result.

Remarks.—*Bacillus coli communis* is one of the most abundant and most characteristic of sewage bacteria. In this and a previous report it has been shown that its number may exceed 100,000 per c.c. of London crude sewage. It has likewise been shown that it survives the processes at work in the biological filters at the Outfall Works. Apart from its function as one of nature's scavengers, the *B. coli* is of great importance from the point of view of the bacterioscopic examination of water. In the first report a photograph (plate I, figure 1) was given, showing that even in so minute an amount as $\frac{1}{10000}$ c.c. of Crossness crude sewage, *B. coli* and closely allied forms were present, and it was pointed out that a bacterial process of great delicacy exists for the detection of pollution of water with minimal quantities of sewage.

It has been asserted that *B. coli* is abundant everywhere, that it multiplies outside the animal body, that it is present in the intestinal contents not only of human beings but of the higher mammals and birds, and that, therefore, its value as an indication of pollution of water of possibly dangerous sort is *nil*. The fact remains that in crude sewage *B. coli* is present in numbers exceeding 100,000 per c.c., and is absent, or present in but few numbers in a corresponding amount of a water free from suspicion of recent pollution. Moreover, if *B. coli* multiplies outside the animal body under favourable conditions, it also loses its vitality under unfavourable conditions, and we have yet to learn that the excrement of healthy, much less of diseased, mammals and birds is altogether harmless to man.

The *Bacillus coli* may be pathogenic, but can hardly be considered pathogenic in the ordinary sense of the term. Its presence serves rather as an index of the possible presence of other and more objectionable kinds of bacteria.

No apology is needed for describing *B. coli* in a report dealing with the bacteriology of sewage. Its prevalence in sewage, its relation to the proper bacterioscopic examination of drinking water, and the important part it plays as one of Nature's scavengers, all make it desirable to record its chief morphological and biological characteristics. This is the more necessary as the published descriptions of this microbe are often incomplete, and in some cases even misleading.

It will be noted that not only the presence of *B. coli* in the various samples of crude sewage and effluents has been determined, but that, as well, a record has been kept of its relative abundance. Until such records are obtained, not only as regards *B. coli*, but as regards many other species of micro-organisms, the usefulness of bacteriology is restricted.

2. BACILLUS MESENTERICUS.

Sewage variety E.	Sewage variety I.
[An aërobic, non-chromogenic, actively motile, slowly liquefying bacillus.]	[An aërobic, non-chromogenic, actively motile, rapidly liquefying bacillus.]

Source.

London crude sewage; gelatine plate cultures heated to 80° C. for 10 minutes.	London crude sewage; gelatine plate cultures heated to 80° C. for 10 minutes.

Morphology.

* Long bacilli, with rounded ends; solitary, in pairs and long chains.	Medium-sized bacilli, with rounded ends; solitary,† in pairs and chains.

Motility.

Actively motile.	Exceedingly rapid movement.

Spore formation.

Readily forms spores.	Readily forms spores.

Flagella.

‡ A multi-flagellated organism.	A multi-flagellated organism. ‖

Temperature.

Grows well at the room temperature, and exceedingly rapidly at 37° C.	Grows very rapidly at the room temperature, and with great rapidity at 37° C.

---

* See fig. 3, Plate I. † See fig. 5, Plate II. ‡ See fig. 4, Plate I. ‖ See fig. 6, Plate II.

Gelatine plate cultures.

The growth is not very rapid. The deep colonies have a somewhat starlike appearance; the superficial colonies appear as bluish-white, delicate granular films, which are almost coli-like in character. These surface colonies are somewhat irregular in shape, and from the wavy, transparent edge irregular processes are given off.

Under a lower power of the microscope the deep colonies show a central darkish yellow spot, from which root-like processes are given off. The surface colonies are transparent, granular, and striated, and from the spreading edge delicate processes are given off, which spread over the surface of the gelatine, forming often curious and intricate patterns. The colonies do not attain a large size, and liquefaction only slowly sets in.

The growth is characteristically rapid. The deep colonies quickly reach the surface, to form large areas of liquefied gelatine, which become saucer-shaped, and are greyish-white in colour and almost translucent.¶ From the edge of these saucer-shaped surface colonies processes may be given off which are almost of the nature of "swarming islands." These almost translucent areas of liquefied gelatine are rendered greyish-white by the presence of innumerable bacteria, and these bacteria may be gathered together in clumps so as to give rise to a mottled appearance. A thin bacterial film tends to form on the surface of the liquefied gelatine. In a very few days the whole plate is completely liquefied.

Under a low power of the microscope the deep colonies are dark in colour and granular. The surface colonies at first may be almost completely translucent and of an irregular, star-shaped form, but soon form large circular areas of liquefied gelatine, in which the extraordinarily rapid movement of the individual bacilli can be clearly watched.

Gelatine "stab" cultures.

The growth at the surface is slow, and resembles the growth of a superficial colony in gelatine plate culture. Growth takes place all the way down the stab, accompanied by slow liquefaction; delicate tuft-like filaments are given off all down the line of the inoculation. These extend more and more deeply into the solid gelatine as growth proceeds. Gradually at the surface liquefaction sets in, showing itself at first merely by a slight central pitting of the bacterial film. Later, the liquefaction spreads, and the delicate details of growth are lost.

Rapid liquefaction takes place nearly if not quite to the foot of the stab and in funnel form.§ Very soon the liquefaction spreads to the walls of the tube and increases rapidly from above downwards as well as from the stab in an outward direction. The liquefied gelatine has a greyish-white translucent appearance, and has also a somewhat flocculent appearance due to aggregation of little masses of bacteria. Very soon the whole contents of the tube are completely liquefied and converted into a greyish-white turbid fluid.

Gelatine "streak" cultures.

A delicate granular film forms on the surface of the gelatine which is of somewhat limited extent. Liquefaction slowly sets in and shows itself as a longitudinal furrow. After some days the film changes its appearance and shows numberless fine processes radiating outwards and upwards from the central line. Tuft-like processes also extend into the solid gelatine. As liquefaction proceeds all the delicate details of growth become lost.

The growth is not characteristic, as the liquefaction is so rapid that the peptonised gelatine runs down the oblique surface to the foot of the tube, and in a few days the whole of the contents are converted into a turbid greyish-white liquid.

Gelatine "shake" cultures.

No gas bubbles are formed; the gelatine is slowly liquefied.

No gas bubbles are formed; the gelatine is rapidly liquefied.

Agar plate cultures.

In agar plate cultures at 37° C. the growth is so rapid and the colonies spread so much laterally, that frequently the whole plate is covered with a granular dirty yellowish-white film in 24 hours. The spreading edge is often broken up into processes of most varied shape, and which frequently form tree-like patterns.

In agar plate cultures at 37° C. the growth is so rapid that it is difficult to obtain discrete colonies. The growth is yellowish-white, granular, and often of unequal thickness. The spreading edge has less tendency to form tree-like patterns than in the case of *B. mesentericus E.*

Agar "stab" cultures.

In 24 hours at 37° C. there is growth all down the stab, and the surface of the medium is covered with a yellowish-white granular layer, which later may become wrinkled.

In 24 hours at 37° C. there is growth all down the stab, and the surface of the medium is covered with a yellowish-white granular layer, which later may become wrinkled.

Agar "streak" cultures.

In 24 hours at 37° C. a dirty yellowish-white film has covered nearly the whole oblique surface. The spreading edge may extend as irregular processes forming leaf-like patterns. Later the film darkens in colour and becomes wrinkled.

In 24 hours at 37° C. a dirty yellowish-white film has covered nearly the whole oblique surface. The spreading edge may be lobed or fissured and of unequal thickness. The film darkens in colour and becomes wrinkled.

Potato cultures.

The growth is extremely characteristic. At 37° C. in 24 hours the whole surface of the potato is covered with a thin yellowish-white film, which has a characteristic folded, creased and wrinkled appearance. There is an appearance also as of blisters: these dry up, leaving deeply wrinkled skins. The substance of the potato takes on a bright pink colour. Later the film becomes thicker, more deeply pitted and wrinkled, and the colour becomes brown.

The growth is extremely characteristic. In 24 hours at 37° C. the whole surface of the potato is covered with a thick greyish-white moist skin, which is thrown into multiple folds, creasings and wrinkles.** The colour rapidly changes from greyish-white to yellow, and then to brown. The growth, if touched with a platinum needle, is found to be held to the potato by a viscous substance which can be drawn out into long threads.

Broth cultures.

The growth at 37° C. is characteristic even in 24 hours. A greyish-white wrinkled film is formed at the surface, and the liquid below is nearly quite clear. Later the film thickens and becomes deeply pitted and wrinkled.

The growth at 37° C. is characteristic even in 24 hours. A greyish-white wrinkled film is formed at the surface, and the liquid below is nearly quite clear. Later the film thickens, and the wrinkled appearance becomes more marked. It acquires a reddish-brown colour, and the liquid below, which remains nearly quite transparent, also takes on a reddish-brown colour.

¶ See fig. 7, Plate II. § See fig. 8, Plate II. ** See fig. 9, Plate III.

Litmus milk cultures.

In 24 hours at 37° C. slight discolouration has taken place. In 48 hours the liquid is of a dirty yellowish-white colour, with a tinge of red; no clot. In 72 hours a weak clot has formed, and the liquid near the surface is semi-transparent. On gently shaking the tube, a pinkish tinge develops. In five days the clot lies at the foot of the tube as a dirty white mass; above this the liquid is transparent and of a faint yellow colour, which on shaking changes to a pink tint.

In 24 hours at 37° C, the bluish-purple colour of the milk is changed to a dirty yellowish-white; no clot. In 48 hours no clot, but the milk is rapidly becoming transparent. In 72 hours the whole of the contents of the tube are semi-transparent and of a dirty yellowish colour. On gently shaking the tube, the liquid assumes a crushed strawberry colour.

Blood serum cultures.

At 37° C., and in less than 24 hours, nearly the whole of the oblique surface of the medium has become covered with a deeply wrinkled skin. Later, slow liquefaction sets in, and in about 16 days the blood serum is completely liquefied.

At 37° C. decided liquefaction takes place in less than 24 hours, and on the surface of the liquid, at the foot of the tube, a wrinkled skin is formed. Later, the blood serum is completely liquefied.

Indol reaction.

No indol is formed.

No indol is formed.

Reduction of nitrates to nitrites.

[Broth 5 per cent., K N O₃ 0·1 per cent., water 94·90 per cent.]

Great reduction of nitrates to nitrites in 24 hours at 37° C.

No reduction of nitrates to nitrites in 24 hours at 37° C.

Remarks.

Resembles very closely, if it is not identical with, *Bacillus mesentericus ruber*.

The micro-organism is constantly present in sewage in the form of spores; 10, 20, 30, or more spores per c.c. of London crude sewage may be found. It is present in considerable number in the effluents from the biological filters. It resembles closely *B. mesentericus vulgatus*, *B. mesentericus fuscus*, and *B. liodermos*, and is perhaps most closely allied to, if it is not identical with, *B. mesentericus vulgatus*.

3.—SEWAGE PROTEUS.

[*An aërobic, non-chromogenic, actively motile, rapidly liquefying bacillus.*]

Source.—Very abundant in London crude sewage; frequently as many as 100,000 per c.c. of crude sewage.

Morphology.*—Small bacilli, with rounded ends; solitary, in pairs, or sometimes in short chains; involution forms may occasionally be seen.

Motility.—Actively motile.

Spore formation.†—No spores are formed.

Flagella.†—Each rod is possessed of a single flagellum.

Temperature.—The original culture grew better at 20° C. than at 37° C.¶

Gelatine plate cultures.‡—In less than 24 hours at 20° C. the surface colonies appear as delicate granular films of irregular shape. In two days the colonies look like "punched out" circles containing liquefied gelatine, and greyish-white bacterial deposit. The masses of bacteria lying in the liquefied gelatine usually give a mottled appearance to the colonies. Viewed under a low power of the microscope the individual bacilli can be made out, and their active movement watched: the colonies appear darkest centrally and at the circumference, and here and there darker spots may be seen in the liquefied and granular-looking gelatine. The colonies are usually exactly circular in shape with well-defined borders, and no "swarming islands" appear to be given off, as in proteus vulgaris. By the third or fourth day the plate is completely liquefied, the gelatine being converted into a turbid greyish-white liquid.

Gelatine "stab" cultures.§—The growth is very characteristic.—In 24 hours at 20° C. liquefaction has occurred all the way down the path of the needle, and minute bubbles of gas may be watched rising through the turbid, greyish-white liquefied gelatine to the surface. In 48 hours the liquefaction is very much more pronounced; numerous bubbles of gas may be seen at the surface, and also bubbles in the solid gelatine. The bacteria collect at the foot of the liquefied portion as a greyish-white deposit. In four or five days the whole of the gelatine is converted into a greyish-white liquid. If such a culture be heated to 80° C. for 20 minutes the bacilli are killed, but if, after cooling, a portion of the fluid be added to another tube containing solid gelatine, the gelatine in this second tube becomes liquefied.

Gelatine "streak" cultures.—The gelatine is liquefied so rapidly that all details of growth are lost.

Gelatine "shake" cultures.—∥ In 24 hours at 20° C. numerous gas bubbles are formed, and the gelatine is liquefied near the surface.

Agar plate cultures.—The growth is not characteristic. The colonies are more or less circular in shape, they are greyish-white in colour, and have a somewhat moist, glistening appearance.

Agar "stab" cultures.—There is growth all down the stab; gas bubbles may form in the medium; on the surface the growth is like a surface colony in an agar plate culture. In old cultures the surface growth becomes of a brownish colour.

Agar "streak" cultures.—The growth is rapid, but not specially characteristic. A greyish-white layer having a moist glistening appearance is formed, which may extend nearly to the walls of the tube.

Potato cultures.—The growth is not characteristic; a slimy, thin, yellowish-white growth appears on the surface of the potato.

Broth cultures.—Abundant diffuse cloudy growth in 24 hours at 20° C.

Litmus milk cultures.—The growth at 20° C. in litmus milk culture may be described as an acid change from an opaque bluish-purple coloured liquid to a semi-transparent reddish-coloured fluid. The redness is chiefly near the tube, but if the tube be shaken the whole of the contents of the tube assume a crushed strawberry tint. If clotting occurs, it is imperfect in character and possibly is dissolved as soon as it is formed.

Blood serum cultures.—Rapid growth, accompanied by liquefaction.

Indol reaction.—Usually no indol is formed in broth cultures, but in one such culture, kept for twelve days, a feeble indol reaction was observed.

Reduction of nitrates.—Rapid reduction of nitrates to nitrites in 24 hours at 20° C. [Broth, 5 per cent., K N O₃ 0·1 per cent., water 94·9 per cent.]

Remarks.—This organism differs from *Proteus vulgaris*, and has little or no resemblance either to *Proteus mirabilis* or to *Proteus zenkeri*. It differs from *Proteus vulgaris* in many respects, for example: *sewage proteus*

---

* See fig. 10, Plate III.   † See fig. 11, Plate III.   ¶ See, however, notes under heading, Remarks.
‡ See fig. 12, Plate III.   § See fig. 13 (b) (c), Plate IV.
∥ Fig. 13 (a), Plate IV.

has but one flagellum, whereas *Proteus vulgaris* is generally stated to be multi-flagellated; sewage proteus has smaller and shorter rods than *Proteus vulgaris*; sewage proteus shows no "swarming islands" unlike *Proteus vulgaris*. This organism has been called sewage proteus owing to its prevalence in crude sewage, and because of its superficial resemblance to *Proteus vulgaris*, and to other members of the proteus group. It is not unlikely that this micro-organism is frequently mistaken for the true *Proteus vulgaris*. It also differs in a number of important respects from *Bacillus proteus urinæ*, an organism isolated from the urine of a patient suffering from cystitis by Dr. Horton Smith, and carefully described by him in the *Journal of Pathology and Bacteriology*, December, 1896.

It must not, however, be supposed that *Proteus vulgaris* is not present in London crude sewage, but that its numbers are not as great as seems to be generally supposed. In figure 1, Plate I., a micro-photograph is shewn of the true proteus, which was isolated from a sample of Crossness raw sewage. The preparation was obtained by making a surface gelatine plate culture of this micro-organism, and then making "impressions" from the "swarming islands" produced by its growth on the surface of the gelatine in 20 hours at 20° C.

A culture of "sewage proteus" proved to be very virulent. Thus, 1 c.c. of a 24 hours' broth culture was injected subcutaneously into a guinea-pig. The animal was found dead on the second day, but had probably died on the preceding day. The organism was recovered from the heart's blood in pure culture. [Subsequent work yielded in some measure anomalous results. Thus, as freshly isolated from sewage, "sewage proteus" sometimes grew well at 37° C., and sometimes in an imperfect manner. Moreover, the cultures were in some cases virulent and in others not.]

4.—BACILLUS FRONDOSUS.

[*An aerobic, non-chromogenic, motile, slowly liquefying bacillus.*]

Source.—London crude sewage; gelatine plate cultures heated to 80° C. for 10 minutes.
Morphology.—Large bacilli, with rounded ends; solitary, in pairs, and in chains of varying length.
Motility.—Motile.
Spore formation.*—Forms spores readily at the room temperature.
Temperature.—No growth at 37° C. Grows fairly rapidly at the room temperature.
Gelatine plate cultures.—The surface colonies appear as white, coarsely granular films of irregular shape, which send out curious processes of varied form. These grotesquely-shaped processes resembles bits of sea-weed flattened out. Liquefaction sets in very slowly and shows itself by a slight pitting near the centre of each colony. The deep colonies are not characteristic. Under a low power of the microscope the superficial colonies have a granular and laminated appearance, and show at the spreading edge processes which are of the most varied shapes, and which often form patterns of great delicacy and beauty.
Gelatine "stab" cultures.—The growth at the surface is like a superficial colony in gelatine plate culture. The growth down the stab is not characteristic. Liquefaction slowly sets in at the surface, and eventually all the delicate details of growth are lost.
Gelatine "streak" cultures.—The growth is like an elongated surface colony in plate culture. Briefly, a white coarsely granular film forms on the sloping surface of the medium which peripherally shoots out processes of very irregular shape. Liquefaction first shows itself as a longitudinal and central pitting of the bacterial film.
Gelatine "shake" cultures.—No bubbles of gas are formed.
Agar plate cultures.—The colonies are white in colour; they do not grow in the same characteristic way as the colonies in gelatine plate culture.
Agar "stab" cultures.—Growth occurs all the way down the stab, and at the surface a greyish white-layer is formed of irregular shape which resembles somewhat a surface colony in gelatine plate culture.
Agar "streak" cultures.—The growth extends over the surface of the medium as a greyish-white film. The growth at the periphery in some measure simulates the corresponding growth on gelatine.
Potato cultures.—The growth is characteristic from a negative point of view as it is of a transparent colourless character.
Broth cultures.—Grows slowly; a white bacterial deposit collects at the foot of the tube, leaving the liquid above fairly clear.
Litmus milk cultures.—In milk at 20° C. No change is visible in three days. Later the milk turns slightly acid, no clotting is visible even after one month.
Blood serum cultures.—At 20° C. a greyish-white, somewhat granular and film-like growth develops in a few days. The edge is irregular. Apparently, no liquefaction takes place.
Indol reaction.—No indol is formed in broth cultures, after 6 days incubation at 20° C.
Reduction of nitrates to nitrites.—No reduction takes place in 4 days at 20° C. [Broth, 5 per cent., K N O, 0·1 per cent., water 94·9 per cent.]
Remarks.—This micro-organism has been compared with the descriptions of all the aerobic, motile, non-chromogenic, liquefying bacteria, and it resembles none of them sufficiently closely to suggest identity. It has been called *B. Frondosus* because the spreading edge of the colonies sometimes have a leafy appearance.

5.—BACILLUS FUSIFORMIS.

[*An aerobic, non-chromogenic, motile, non-liquefying bacillus.*]

Source.—London crude sewage; gelatine plate cultures heated to 80° C. for ten minutes.
Morphology.—Large bacilli, with rounded ends; solitary, in pairs, and in chains.
Motility.—Motile.
Spore formation.†—Forms spores at the room temperature. These are large, and give a spindle-shaped appearance to the cells.
Temperature.—No growth at 37° C. Grows slowly at the room temperature.
Gelatine plate cultures.—The growth is very slow. The surface colonies are circular, of an opaque porcelain-white, glistening appearance. The periphery is slightly transparent. By transmitted light the colonies are yellowish in colour. In old cultures the white colour takes on a yellowish tint. The deep colonies are not characteristic. Microscopically, under a low power, no delicate details of growth can be made out. No liquefaction of the gelatine takes place.
Gelatine "stab" cultures.—Tardy growth, no liquefaction of the gelatine. A white growth appears along the line of the stab, and on the surface a scanty yellowish-white layer very slowly develops.
Gelatine "streak" cultures.—A glistening porcelain-white layer is slowly formed, which is opaque, except at the margins, which are slightly transparent. By transmitted light and by reflected light in old cultures a slight yellowish tint may be seen. The growth does not extend far from the actual line of inoculation. No liquefaction occurs.
Gelatine "shake" cultures.—No gas bubbles are formed.
Agar plate cultures.—The colonies are white and more or less circular in shape; the growth is not characteristic.
Agar "stab" cultures.—The growth is slow and imperfect. A whitish line appears along the line of the stab, and on the surface a thin white layer slowly develops.
Agar "streak" cultures.—A thin whitish layer is slowly formed; the growth is not characteristic.

---

* See fig. 14, Plate IV.   † See fig. 15, Plate IV.   ‡ See fig. 16, Plate IV.

Potato cultures.—A porcelain-white growth slowly developes; afterwards the colour becomes dirty yellowish-white and the bacterial layer becomes unequally thickened.
Broth cultures.—The growth is very scanty and not characteristic.
Litmus milk cultures.—No clotting occurs, and the change in the medium is hardly visible, even after 16 days. Apparently a feeble acidity of the milk results.
Blood serum cultures.—Little or no growth on the surface, even after the lapse of some time; some growth, however, takes place in the fluid at the foot of the tube.
Indol reaction.—No indol is formed in broth cultures.
Reduction of nitrates to nitrites.—Negative after 12 days at 20° C. [Broth 5 per cent., KNO_3 0·1 per cent., water 94·9 per cent.]
Remarks.—This micro-organism has a somewhat negative character of growth in all the nutrient media ordinarily in use. It has been called *B. fusiformis* owing to the shape of the spores, spindle-shaped. So far as could be ascertained it belongs to a new species.

6.—BACILLUS SUBTILISSIMUS.

[*An aërobic, non-chromogenic, non-motile, non-liquefying bacillus (? micro-coccus).*]

Source.—Crude sewage.
Morphology.*—In most cultures it appears as a large micro-coccus, but if "impression" preparations be made from surface colonies in a gelatine plate, it will be seen that at the spreading edge the elements are distinctly longer than broad, nearer the centre they are oval and frequently united in pairs, and at or about the centre they are perfectly spherical.
Motility.—No motility has been observed even in recent broth cultures.
Spore-formation.—None.
Temperature.—Does not grow at 37° C., but grows with extreme rapidity at 20° C.
Gelatine plate cultures.—The deep colonies are not characteristic, either on naked eye examination when viewed with a hand-lens, or when examined under a low power of the microscope. The surface colonies are peculiarly characteristic, and grow so rapidly and extend so widely, that a single colony may cover nearly a whole plate in two days. The growth is film-like in character and extremely thin and transparent. It is dull grey in colour and very faintly granular. When viewed under a low power of the microscope, the appearance is not unlike *B. coli*, but the details of the growth are so much more delicate that it is difficult to perceive the slight granulation, faint creasing and delicate veining of the bacterial film. The surface colonies are usually of a more or less circular shape, but the spreading edge is nearly always markedly irregular.
Gelatine "*streak*" *cultures.*†—The growth is like an elongated surface colony. In less than 24 hours a delicate film has spread nearly to the walls of the tube. The spreading edge is very irregular, and in older cultures may present a terraced appearance.
Gelatine "*shake*" *cultures.*—No gas bubbles are formed in the gelatine.
Agar "*streak*" *cultures.*—A white film is formed on the surface having a markedly irregular edge. The lateral expansion is less than in the case of gelatine cultures.
Broth cultures.—Uniform turbidity occurs in 24 hours at 20° C., and a very faint scum forms on the surface. No motility could be made out.
Blood serum.—A thin film forms of a faint yellowish-white colour. No liquefaction occurs.
Litmus milk cultures.—No visible change in 48 hours at 20° C. Later the purple colour of the milk changes to a reddish purple, and later still to a buff colour. On shaking the tube gently the liquid does not assume a red tint. There is no clotting produced and no transparency of the medium occurs even in old cultures.
Indol.—No indol is formed even after 20 days at 20° C.
Remarks.—This microbe has been called *B. subtilissimus* on account of the thin almost gauze-like character of the surface colonies in gelatine plate culture. No micro-organism hitherto described appears to correspond with the above.

7.—BACILLUS SUBTILIS.

SEWAGE VARIETY A.	SEWAGE VARIETY B.
[*An aërobic, non-chromogenic, rapidly liquefying, spore-forming, motile bacillus.*]	[*An aërobic non-chromogenic, rapidly liquefying, spore-forming, motile bacillus.*]
Source.	
Crude sewage and effluents from coke-beds.	Crude sewage and effluents from coke-beds.
Optimum temperature.	
Grows luxuriantly at 37° C., also at the room temperature.	Little or no growth at 37° C., grows luxuriantly at 20° C.
Morphology.	
Large and long bacilli with rounded ends, frequently associated in long chains.	Large and long bacilli with rounded ends, frequently associated in long chains.
Spore formation.	
Forms spores readily.	Forms spores readily.
Motility.	
Waddling sluggish movement.	Waddling sluggish movement.
Gelatine plate cultures.	
Forms in two days at 20° C. circular greyish-white areas of liquid gelatine which rapidly increase in diameter. The white masses of bacteria lying in the liquefied gelatine may present a mottled appearance, or may be arranged in stellate fashion. Under a low power of the microscope (about 80 diam.) the individual bacilli can be clearly seen in the more liquid portion, and their movements watched. At the edge of the colonies the bacilli bore side by side into the non-liquefied gelatine in a highly characteristic way.‡	Rapidly froms greyish-white circles of liquefied gelatine. The masses of bacteria lying in the liquefied gelatine may present a rosette or star-shaped or radiating appearance. Under a low power of the microscope (about 80 diam.) the movement of the individual bacilli can be clearly seen. The parallel arrangement of the bacilli at the periphery of the colonies is absent, or not so well marked as in the case of variety A. A skin forms on the surface of the liquefied gelatine. The growth is not so rapid as in the case of variety A. Sometimes the colonies are not exactly circular in shape.

---

* See fig. 17, Plate V. † Fig. 18, Plate V. ‡ Fig. B, plate VII.

Agar "streak" cultures.

In two days, at 37° C., an abundant creamy white layer, covering nearly the whole surface, which on close inspection shows numerous minute circular areas where the growth, instead of being opaque, is semi-transparent.

In 20 hours, at 20° C., a greyish-white layer, not specially characteristic. In 48 hours, growth somewhat dry and granular looking. In four days, curious wrinkled appearances. Ridges, formed by the unequal rate of growth, or by the contraction of the bacterial skin, stand out from the surface of the medium about one-sixteenth of an inch, and usually are arranged in more or less transverse folds.*

Gelatine "stab" cultures.

In 2 days, at 20° C., liquefaction has occurred right down to the foot of the stab. White flocculent masses of bacteria sink through the liquefied and grey-coloured gelatine to the foot of the stab. The liquefaction extends rapidly from the line of inoculation outwards as well as from the surface downwards. A scum forms on the surface, but no distinct skin is formed.†

Rapid liquefaction all the way down the stab, but as the growth proceeds the liquefaction spreads in cylindrical fashion from above downwards rather than from within outwards from the region of the line of inoculation. A distinct skin forms on the surface, which eventually sinks in the liquefied gelatine.‡

Liquefaction.

Rapidly liquefies gelatine and blood serum.

Liquefies gelatine and blood serum fairly rapidly.

Gas formation.

Forms no gas in gelatine "shake" cultures.

Forms no gas in gelatine "shake" cultures.

Broth cultures.

Diffuse cloudiness in 24 hours at 37° C. A scum forms on the surface which readily falls to the foot of the tube on shaking.

Diffuse cloudiness: a skin forms on the surface which is brittle and sinks on shaking the tube.

Indol.

Forms no indol.

Forms no indol.

Litmus milk cultures.

In 2 days, at 37° C., there is complete discoloration and a clot has formed; no redness is visible. In 5 days a white clot occupies about one-third of the bulk of medium, which appears to be slowly peptonised. The liquid surrounding the clot is pale yellow in colour, and semi-transparent without pink coloration.

The purple-blue colour of the litmus milk gradually fades, but no clotting or redness occurs. In 8 days the milk is almost transparent and of a pale dirty yellow colour; on shaking the tube the liquid assumes a faint pink tinge. By the 21st day the milk has changed to a semi-transparent dirty yellow liquid; on shaking the tube the contents assume a reddish tint.

Potato cultures.

In 2 days, at 37° C., a dirty-white layer is formed with a yellowish tint.

A white, rather dry-looking coat is formed. Later portions of the growth become upraised, and sometimes present a worm-like appearance.

Blood serum cultures.

In 24 hours, at 37°, a white layer develops, accompanied with liquefaction of the medium.

In 20 hours, at 20° C., a thin whitish layer is formed. In 48 hours commencing liquefaction and markedly wrinkled skin.‖

Remarks.—Several varieties of *B. subtilis* occur in sewage. For the purpose of description I have named these two :—Sewage variety A, and sewage variety B. The fact, however, of the latter organism not growing at 37° C. makes it doubtful whether it should be considered a variety of *B. subtilis*.

8.—BACILLUS MEMBRANEUS PATULUS.

[*An aërobic, non-chromogenic, slowly liquefying, spore-forming (?), non-motile bacillus.*]

Source.—Crude sewage and effluent from coke-beds.
Optimum temperature.—Grows well at 37° C., and at the room temperature.
Morphology.—A very large bacillus which forms long chains.§
Spore formation.—Old culture resist heating to 80° C. for ten minutes, and the microbe has been found in cultures made from sewage which had previously been heated to 80° C. for ten minutes. A satisfactory double-stained preparation, however, has not been obtained.
Motility.—No motility could be made out, even in a 20 hours' broth culture.
Gelatine plates.—The surface colonies appear as coarsely granular greyish-white films of somewhat irregular shape. From the spreading edge of the colonies processes extend in a tortuous fashion over the surface of the medium often forming patterns of great delicacy and beauty. Beneath the surface film-like growth, slow liquefaction of the gelatine occurs. The growth is rapid. Under a low power of the microscope the colonies present a characteristic granular and striated appearance.¶
Oblique gelatine cultures.—In less than two days, at 20° C., a coarsely granular film forms on the surface of the gelatine which spreads rather rapidly and may extend nearly to the walls of the tube. From the spreading edge processes are given off which wind over the surface of the gelatine in a characteristic way. Soon a longitudinal pitting of the bacterial film along the line of inoculation is observed, and later on as liquefaction proceeds, the growth slips down to the foot of the tube, and all delicate details are lost. The growth is like an elongated surface colony in gelatine plate culture.**
Oblique Agar cultures.—In one night, at 37° C., the growth appears as a coarsely granular, semi-transparent, greyish-white film. By the second day the growth is less transparent and less granular looking. In old cultures the growth assumes a tuberculated appearance.
Gelatine "stab" cultures.††—The growth varies, as sometimes there is liquefaction down the line of the stab with tuft-like processes extending into the solid medium, and at other times there is almost no liquefaction along the line of inoculation, and the processes extend nearly to the walls of the tube, giving rise to an appearance of great beauty. The growth on the surface is like the growth of a surface colony in a gelatine plate culture.
Liquefaction.—Liquefies gelatine, but not rapidly. Produces only very slight liquefaction of blood serum even in cultures kept at 37° C. for 16 days.
Gas formation.—Forms no gas in gelatine "shake" cultivations.

---

* Fig. 20, Plate V. † Fig. A, plate VII. ‡ Fig. C., plate VII., and Fig. 19, plate V. ‖ Fig. D., plate VII.
§ Fig. 21, plate VI. ¶ Fig. F, plate VIII. ** Fig. 22, plate VI., and Fig. E, plate VIII. †† Fig. 23, Plate VI.

Broth cultures.—Grows very rapidly at 37° C. The cloudiness throughout the medium is flocculent rather than diffuse. An abundant white bacterial deposit collects at the foot of the tube. On the surface a skin is formed, which sinks on shaking the tube, but is re-formed in one night.
Indol formation.—Forms no indol in broth cultures.
Litmus milk cultures.—In 24 hours, at 37° C., the milk has become slightly discoloured; later, a weak gelatinous clot is formed, and the medium turns faintly acid.
Potato cultures.—The growth is not characteristic; a dirty, faint yellowish-grey coloured growth in 24 hours at 37° C.
Blood serum cultures.—A granular greyish-white film is formed. Only very slight liquefaction occurs, even in old cultures.
Reduction of nitrates.—Great reduction of nitrates to nitrites in one night at 37° C. (Bouillon 5 per cent. KNO_3, 1 per cent.)
Remarks.—This organism does not appear to resemble at all closely the descriptions of any of the bacteria found in sewage and elsewhere. On account of its spreading, film-like character of growth, it has been termed *B. membraneus potulus.*

9.—BACILLUS CAPILLARIS.

[*An aerobic, non-chromogenic, rapidly liquefying, spore-forming (?), motile bacillus.*]

Source.—Crude sewage and effluents from coke beds.
Optimum temperature.—Grows luxuriantly at 37° C., also at room temperature.
Morphology.—A large bacillus, forming long chains.
Spore formation.—Old cultures resist heating to 80° C., and this microbe commonly occurs in cultivations made from sewage which have previously been heated to 80° C. for 10 minutes. A satisfactory double-stained preparation, however, has not been obtained.
Motility.—This organism is motile.
Gelatine plate cultures.—The colonies in the depth have a characteristic fluffy appearance. They rapidly increase in size, and reaching the surface quickly liquefy the gelatine. The growth is filamentous in the depth, and on the surface from the spreading edge of the colonies, delicate film-like processes are given off, which extend over the surface of the medium to form irregular patterns. Later, the finer details of growth are lost, owing to the rapid liquefaction of the gelatine, the colonies eventually appearing as large, more or less circular, areas of liquefied gelatine with greyish-white contents.
Agar "streak" cultures.—In 24 hours an opaque-white growth of limited extent appears, which along the spreading edge is slightly transparent and granular-looking. Later, the growth may extend laterally to cover a wide extent of surface.
Gelatine "streak" cultures.—A longitudinal furrow appears due to the liquefaction of the gelatine, but from the edge of the furrow delicate processes may be seen extending in irregular fashion over the surface of the solid gelatine. The furrow is nearly clear as the bacteria slip down the sloping surface with the liquid gelatine, and collect at the foot of the tube. Soon all details of growth are lost, owing to the progressive liquefaction of the medium.
Gelatine "stab" cultures.—Liquefaction takes place in funnel form, and extends down the stab to an extent varying in different cultures. Along the line of the inoculation feathery processes are given off, which extend into the solid gelatine for a short distance. The masses of bacteria gradually sink to the foot of the liquefied medium.
Liquefaction.—Liquefies gelatine rapidly, blood serum less rapidly.
Gas formation.—No gas is formed in gelatine "shake" cultures. Sometimes the colonies in "shake" cultures have a beautiful star-shaped appearance, but at other times they are globular.
Broth cultures.—In broth at 37° C. there is diffuse cloudiness with flocculent masses scattered throughout the medium. Later, a skin forms on the surface, and an abundant white bacterial deposit collects at the foot of the tube, leaving the liquid above fairly clear.
Indol.—No indol is formed in broth cultures.
Litmus milk cultures.—No decided change, even after 72 hours at 37° C. In five days weak gelatinous clot, but no apparent acidity.
Potato cultures.—By the second day at 37° C. a fairly abundant creamy coating has formed on the surface of the medium, and later this becomes of a dirty-brown colour.
Blood serum cultures.—By the second day, at 37° C., a white-coloured growth, having no special characters. Liquefaction slowly sets in, and by the 16th day is practically complete.
Reduction of nitrates.—Reduces nitrates to nitrites.
Remarks.—This microbe has been called *B. capillareus* because of the hair-like character of its growth in gelatine plate cultures. It resembles in some respects *B. mycoides*, *B. subtilis* and *B. vessenterius*, but not sufficiently to suggest a similarity of species.

VI.—DESCRIPTION OF MICRO-PHOTOGRAPHS AND DIAGRAMMATIC DRAWINGS ACCOMPANYING REPORT.

Fig. 1.—*Proteus vulgaris*; impression preparation from "swarming islands" on gelatine; 20 hours' growth at 20° C. × 3,000.

Fig. 2.—*B. enteritidis sporogenes* (Klein); microscopic double-stained preparation from a serum culture, showing spores. × 2,000.

Fig. 3.—*B. mesentericus* sewage variety E; microscopic preparation from a 20 hours' agar culture at 20° C. × 1,000.

Fig. 4.—*B. mesentericus* sewage variety E; microscopic preparation stained by V. Ermengem's method and showing numerous flagella, from a 20 hours' agar culture at 20° C. × 1,000.

Fig. 5.—*B. mesentericus* sewage variety I; microscopic preparation from a 20 hours' agar culture at 20° C. × 1,000.

Fig. 6.—*B. mesentericus* sewage variety I; microscopic preparation stained by V. Ermengem's method, showing numerous flagella; from a 20 hours' agar culture at 20° C. × 1,000.

Fig. 7.—*B. mesentericus* sewage variety I; gelatine plate culture. About natural size.

Fig. 8.—*B. mesentericus* sewage variety I; gelatine "stab" cultures. From left to right—three days, two days, one day's growth at 20° C. About natural size.

Fig. 9.—*B. mesentericus* sewage variety I; potato culture, one day's growth at 37° C. Very slightly enlarged.

Fig. 10.—"*Sewage proteus.*" Microscopic preparation from an agar culture; 24 hours' growth at 20° C. × 1,000.

Fig. 11.—"*Sewage proteus.*" Microscopic preparation stained by V. Ermengem's method, showing a single flagellum at one end of each rod; from a 24 hours' growth agar culture at 20° C. × 1,000.

* Fig. 24, plate VI.   † Fig. G, plate VIII.

Fig. 12.—"*Sewage proteus*", gelatine plate culture, two days' growth at 20° C. About natural size.
Fig. 13.—"*Sewage proteus.*" From left to right—gelatine "shake" culture, 24 hours at 20° C.; gelatine "stab" culture, 24 hours' growth at 20° C.; gelatine "stab" culture, 48 hours' growth at 20° C. About natural size.
Fig. 14.—*B. frondosus*. Microscopic double-stained preparation from an agar culture showing spores. × 1,000.
Fig. 15.—*B. frondosus*. Gelatine plate culture. About natural size.
Fig. 16.—*B. fusiformis*. Microscopic double-stained preparation, showing spores. × 1,000.
Fig. 17.—*B. subtilissimus*. Impression preparation from a gelatine plate culture. × 1,000.
Fig. 18.—*B. subtilissimus*. Gelatine "streak" cultures, 24 hours' growth at 20° C. Natural size.
Fig. 19.—*B. subtilis*. Sewage variety B. gelatine "stab" cultures. 1 day's and 3 days' growth at 20° C. About natural size.
Fig. 20.—*B. subtilis*. Sewage variety B. Oblique agar culture, 3 days' growth at 20° C. About natural size.
Fig. 21.—*B. membraneus patulus*. Impression preparation from gelatine plate culture. × 1,000.
Fig. 22.—*B. membraneus patulus*. Oblique gelatine culture. About natural size.
Fig. 23.—*B. membraneus patulus*. Gelatine "stab" culture, 3 days' growth at 20° C. About natural size.
Fig. 24.—*B. capillareus*. Impression preparation from a gelatine plate culture, 20 hours' growth at 20° C. × 1,000.
Fig. A.—*B. subtilis*, sewage variety A. Gelatine "stab" culture—
 (a) Two days' growth at 20° C.
 (b) Three days' growth at 20° C.
 (c) Four days' growth at 20° C.
Fig. B.—*B. subtilis*, sewage variety A. Colony in gelatine plate, under low power of microscope—Two days' growth at 20° C.
Fig. C.—*B. subtilis*, sewage variety B. Gelatine "stab" cultures—
 (a) Two days' growth at 20° C.
 (b) The same as (a) but at a later stage.
Fig. D.—*B. subtilis*, sewage variety B. Blood serum culture—48 hours' growth at 20° C.
Fig. E.—*B. membraneus patulus*. Oblique gelatine culture.
Fig. F.—*B. membraneus patulus*. Showing the appearance, under a low power of the microscope, of the delicate film-like processes which extend over the surface of the medium in gelatine plate cultures.
Fig. G.—*B capillareus*. Under a low power of the microscope.
 (a) Deep-seated colony in gelatine, 20 hours' growth at 20° C.
 (b) Colony partly deep and partly superficial, 20 hours' growth at 20° C.
 (c) Colony at a later stage of growth, and after liquefaction had set in.

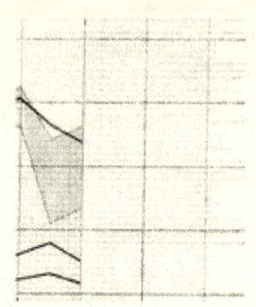

VII.—FURTHER BACTERIOLOGICAL RECORDS FROM AUGUST 9th TO DECE[MBER]

APPENDIX A.—Showing the total number of bacteria in 1 c.c. of Crossness crude sewage; and in 1 c.c. of effluents from 4-ft., 6-[ft...]

Total number of bacteria in 1 c.c.

Date	Crossness crude sewage	Effluent from 4-ft. Coke-bed	Effluent from 6-ft. (primary) Coke-bed	Effluent from 6-ft. (secondary) Coke-bed
1898.				
August 19	5,800,000	3,400,000	—	—
„ 24	4,100,000	—	5,700,000	—
September 14	8,000,000	3,400,000	—	—
„ 21	8,800,000	—	7,200,000	—
„ 28	7,500,000	7,200,000	—	—
October 5	10,500,000	—	6,000,000	—
„ 12	4,000,000	4,500,000	—	—
„ 21	8,000,000	—	15,800,000	—
„ 26	3,200,000	—	—	3,100,000
November 9	7,000,000	8,800,000	—	—
„ 16	3,800,000	—	5,300,000	—
„ 23	4,800,000	—	—	4,500,000
„ 30	13,500,000	5,400,000	—	—
December 7	5,600,000	—	1,000,000	—
„ 14	19,500,000	—	—	3,250,000
„ 21	7,400,000	6,700,000	—	—

...AL RECORDS FROM AUGUST 9th TO DECEMBER 31st, 1898.

Crossness crude sewage; and in 1 c.c. of effluents from 4-ft., 6-ft. (primary), and 6-ft. (secondary) Coke-bed.

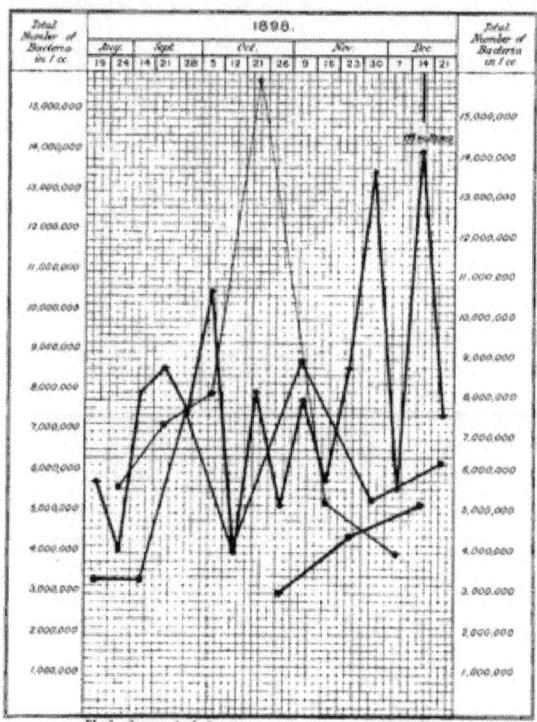

Black = Crossness crude sewage
Red = Filtrate from 4 ft. Coke-bed
Blue = Filtrate from 6 ft. (primary) Coke-bed
Green = Filtrate from 6 ft. (secondary) Coke-bed

Addendum B, showing the number of bacteria causing liquefaction of gelatine in 1 cc. of Crossness crude sewage; and in 1 cc. of effluents

Number of bacteria causing liquefaction of gelatine in 1 cc.—

Date	Crossness crude sewage	Effluent from 4-foot coke-bed	Effluent from 6-foot (primary) Coke-bed	Effluent from 6-foot (secondary) Coke-bed
1898.				
August 19th	600,000	400,000	—	—
„ 24th	600,000	—	1,100,000	—
September 14th	700,000	700,000	—	—
„ 21st	1,200,000	—	1,200,000	—
„ 28th	1,200,000	1,200,000	—	—
October 5th	2,400,000	—	600,000	—
„ 12th	700,000	1,200,000	—	—
„ 21st	1,000,000	—	1,000,000	—
„ 26th	700,000	—	—	500,000
November 9th	1,900,000	1,400,000	—	—
„ 16th	1,600,000	—	1,500,000	—
„ 23rd	900,000	—	—	400,000
„ 30th	900,000	500,000	—	—
December 7th	1,400,000	—	900,000	—
„ 14th	1,400,000	—	—	1,400,000
„ 21st	2,200,000	800,000	—	—

crude sewage; and in 1 cc. of effluents from 4-foot, 6-foot (primary), and 6-foot (secondary) Coke beds.

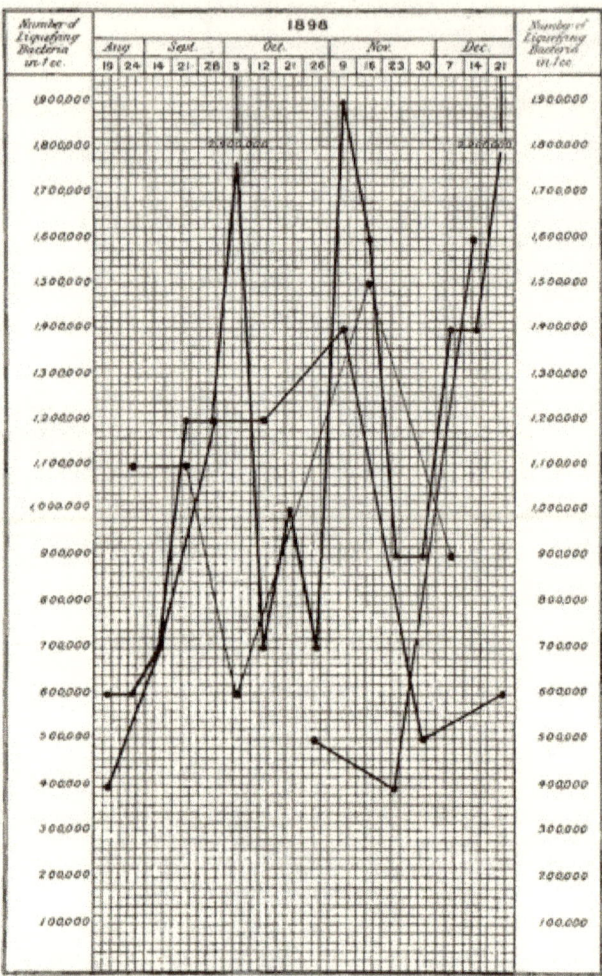

Black = Crossness Crude Sewage.
Red = Filtrate from 4 ft. Coke bed.
Blue = Filtrate from 6 ft. (primary) Coke bed.
Green = Filtrate from 6 ft. (Secondary) Coke bed.

APPENDIX C.—Showing the number of spores of bacteria in 1 c.c. of Crossness crude sewage ; and in 1 c.c. of effluent

NUMBER OF SPORES OF BACTERIA IN 1 C.C.

Date.	Crossness crude sewage.	Effluent from 4-ft. Coke bed.	Effluent from 8-ft. (primary) Coke bed.	Effluent from 8-ft. (secondary) Coke bed.
1898. August 19	220	200	—	—
„ 24	170	—	180	—
September 14	*	*	—	—
„ 21	250	—	190	—
„ 28	260	490	—	—
October 5	200	—	130	—
„ 12	220	240	—	—
„ 21	340	—	430	—
„ 26	200	—	—	220
November 9	150	140	—	—
„ 16	240	—	180	—
„ 23	510	—	—	320
„ 30	440	310	—	—
December 7	70	—	170	—
„ 14	800	—	—	420
„ 21	270	140	—	—

* The rapid liquefaction of the gelatine prevented accurate counting.

in crude sewage; and in 1 c.c. of effluent from 4-ft., 6-ft. (primary), and 6-ft. (secondary) Coke-bed.

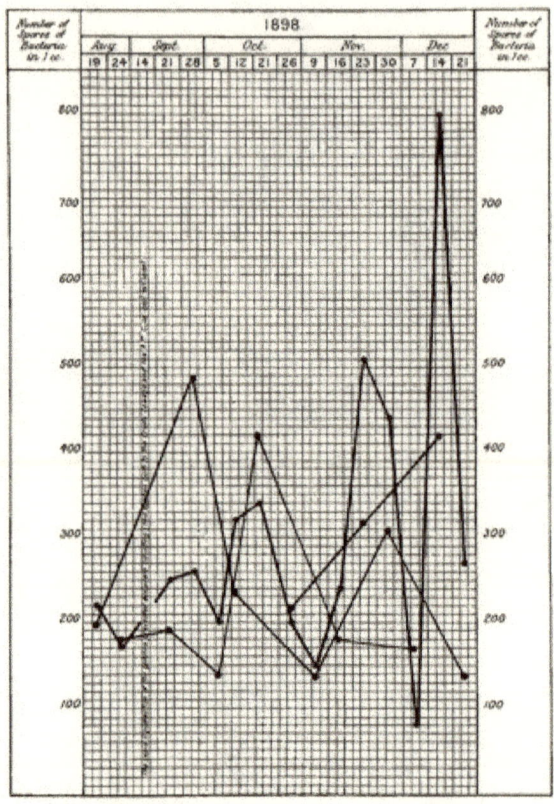

Black = Ooziness Crude Sewage.
Red = Filtrate from 4ft Coke-bed.
Blue = Filtrate from 6ft (primary) Coke-bed.
Green = Filtrate from 6ft (secondary) Coke bed.

APPENDIX D.—Showing the number of B. coli (or closely allied forms) in 1 c.c. of Claremeres crude sewage (16 samples), in 1 c.c. of from 6-ft. (primary) coke-bed (6 samples), and in 1 c.c. of effluent from 6-ft. (secondary) coke-bed (2 samples). [Calculated from the number plate cultures containing 0·00001 c.c. of sample.]

NUMBER OF B. COLI (OR CLOSELY ALLIED FORMS) IN 1 C.C.

Date.		Crude sewage.	Effluent from 4-ft. coke-bed.	Effluent from 6-ft. primary coke-bed.	Effluent from 6-ft. secondary coke-bed.
1898.					
August	19	100,000	None in 0·00001 c.c.	—	—
,,	24	300,000	—	400,000	—
September	14	600,000	400,000	—	—
,,	21	1,600,000	—	700,000	—
,,	28	1,000,000	900,000	—	—
October	5	1,200,000	—	1,300,000	—
,,	12	800,000	500,000	—	—
,,	21	800,000	—	800,000	—
,,	26	None in 0·00001 c.c.	—	—	None in 0·00001 c.c.
November	9	600,000	600,000	—	—
,,	16	300,000	—	500,000	—
,,	23	600,000	—	—	100,000
,,	30	600,000	200,000	—	—
December	7	600,000	—	200,000	—
,,	14	600,000	—	—	100,000
,,	21	500,000	600,000	—	—

c.c. of Crossness crude sewage (16 samples); in 1 c.c. of effluent from 4-ft. coke-bed (7 samples); in 1 c.c. of effluent dry) coke-bed (2 samples). [Calculated from the number of colonies indistinguishable from B. coli in phenol gelatine

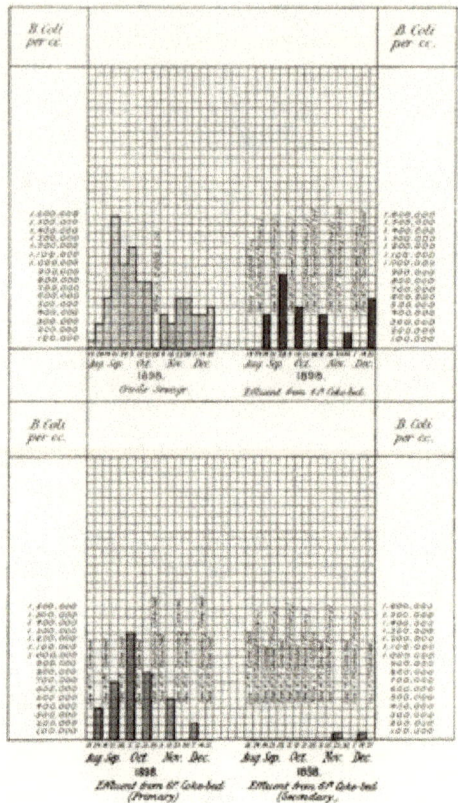

ANNEXURE to E.—Showing the number of spores of B. enteritidis sporogenes (Klein) in 1 c.c. of Crossness crude sewage (17 samples); in effluent from 6-ft. (primary) coke-bed (17 samples); in 1 c.c. of effluent from laboratory vessel (effluent from 6-ft. coke-bed again treated in tubes coke-bed) (5 samples).

NUMBER OF SPORES OF B. ENTERITIDIS SPOROGENES (KLEIN) IN 1 C.C.

Date	Crude sewage	Effluent from 6-ft. coke-bed.	Effluent from 6-ft. primary coke-bed.	Effluent from laboratory vessel (6-ft. coke-bed effluent again treated).	Effluent from 5-ft. secondary coke-bed.
1908 Aug. 19	+0·1 & 0·01	+0·1 & 0·01 c.c.	+0·1	+0·1	—
,, 24	+0·1 & 0·01; −0·001 c.c.	+0·1 & 0·01 c.c.	+0·1, 0·01; −0·001 c.c.	+0·1; −0·01 c.c.	—
Sept. 14	+0·1, 0·01 & 0·001 c.c.	+0·1, 0·01; −0·001 c.c.	+0·1; −0·01 c.c.	—	+0·1 & 0·01 c.c.
,, 21	+0·1, 0·01 & 0·001 c.c.	+0·1 & 0·01 c.c.	+0·1, 0·01; −0·001 c.c.	—	+0·1 & 0·01 c.c.
,, 28	+0·1, 0·01; −0·001 c.c.	+0·1, 0·01; −0·001 c.c.	+0·1 & 0·01 c.c.	—	+0·1 & 0·01 c.c.
Oct. 5	+0·1, 0·01; −0·001 c.c.	+0·1 & 0·01 c.c.	+0·1, 0·01; −0·001 c.c.	—	+0·1 & 0·01 c.c.
,, 12	+0·1, 0·01 & 0·001 c.c.	+0·1, 0·01 & 0·001 c.c.	+0·1 & 0·01 c.c.	—	+0·1; −0·01 c.c.
,, 19	+0·1, 0·01; −0·001 c.c.	+0·1 & 0·01 c.c.	+0·1, 0·01; −0·001 c.c.	—	+0·1 & 0·01 c.c.
,, 26	+0·1, 0·01; −0·001 c.c.	−0·1 & 0·01 c.c.	+0·1 & 0·01 c.c.	—	+0·1, 0·01 & 0·001 c.c.
Nov. 2	+0·1 c.c.	+0·1 c.c.	+0·1 c.c.	—	+0·1 c.c.
,, 9	+0·1, 0·01 & 0·001 c.c.	+0·1, 0·01 & 0·001 c.c.	+0·1 & 0·01 c.c.	—	+0·1 & 0·01 c.c.
,, 16	+0·1, 0·01 & 0·001 c.c.	+0·1 & 0·01 c.c.	+0·1, 0·01; −0·001 c.c.	—	+0·1; −0·01 c.c.
,, 23	+0·1, 0·01 & 0·001 c.c.	+0·1 & 0·01 c.c.	+0·1 & 0·01 c.c.	—	+0·1, 0·01; −0·001 c.c.
,, 30	+0·1, 0·01 & 0·001 c.c.	+0·1, 0·01; −0·001 c.c.	+0·1 & 0·01 c.c.	—	+0·1 & 0·01 c.c.
Dec. 7	+0·1; −0·01 & 0·001 c.c.	−0·1 & 0·01 c.c.	+0·1, 0·01; −0·001 c.c.	—	+0·1; −0·01 c.c.
,, 14	+0·1, 0·01; −0·001 c.c.	+0·1 & 0·01 c.c.	+0·1 & 0·01 c.c.	—	+0·1, 0·01; −0·001 c.c.
,, 21	+0·1, 0·01 & 0·001 c.c.	+0·1, 0·01 & 0·001 c.c.	+0·1; −0·01 c.c.	—	+0·1; −0·01 c.c.

The sign + signifies the presence and the sign − the absence of sperm of B. enteritidis sporogenes (Klein).

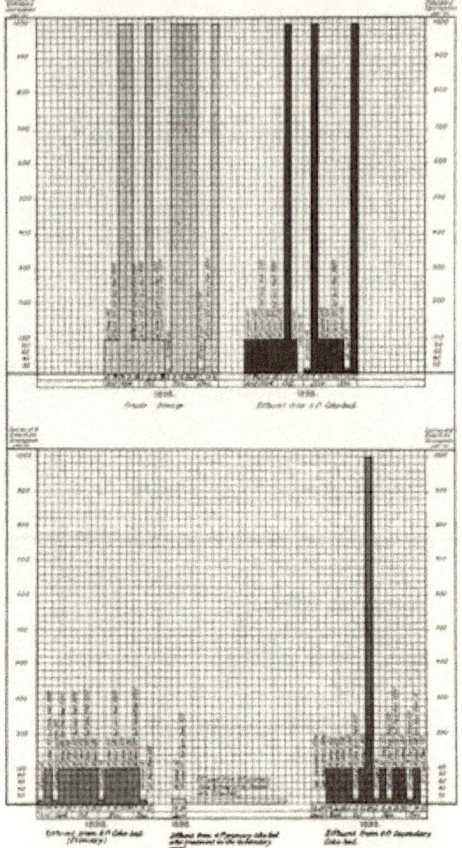

Diagram 2—Percentage purification of crude sewage by coke-beds. These percentages have been calculated from the oxygen absorbed in four hours from permanganate by the raw sewage and by the effluent respectively. Daily averages—

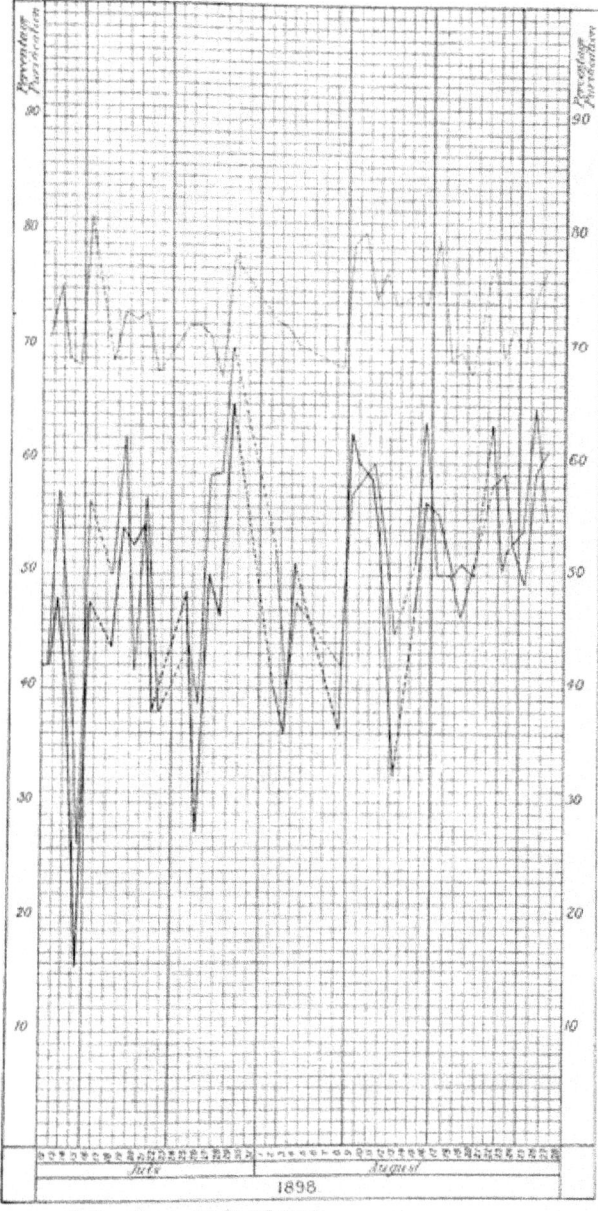

——— Continuous line represents Coke-bed at work.
------ Broken line represents Coke-bed resting and not at work.
Black - Single Coke-bed.
Red - Primary Coke-bed.
Green - Secondary Coke-bed.

Diagram 3, showing the total number of bacteria in 1 cc. of Crossness crude sewage (ten samples); in 1 cc. of effluent from 4-foot coke-bed (eight samples); and in 1 cc. of effluent from 6-foot coke-bed (two samples).

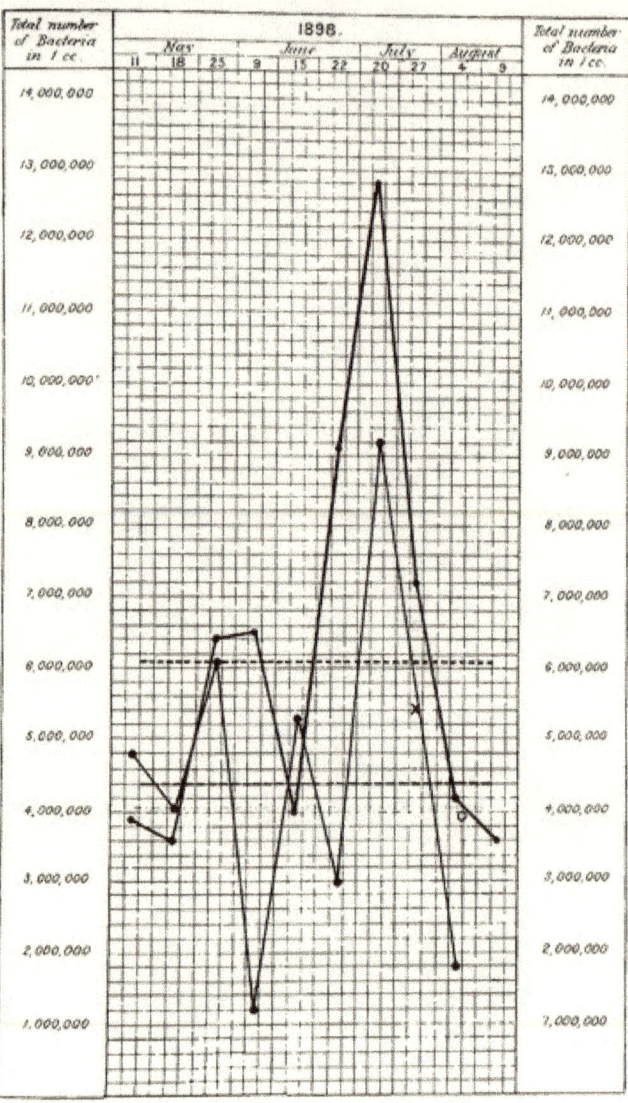

Black = Crossness Crude Sewage (- - - - = average of 10 Expts.)
Red. = Effluent from 4ft Coke-bed (- - - = average of 8 Expts.)

X = No record for this date (4ft Coke bed effluent.)
O = " " " " (6ft Coke bed. effluent)

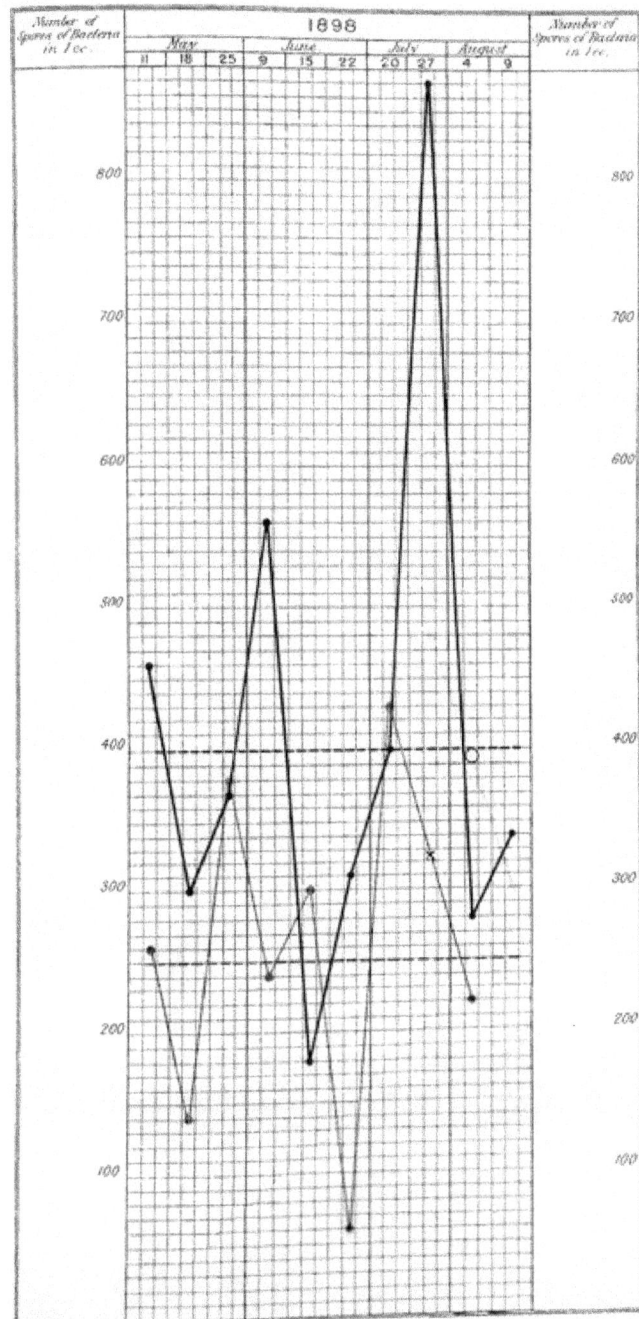

Diagram 4, showing the number of spores of bacteria in 1 cc. of Crossness crude sewage (10 samples); in 1 cc. of effluent from 4-foot coke-bed (8 samples); and in 1 cc. of effluent from 6-foot coke-bed (2 samples).

Black = Crossness Crude Sewage (------ average of 10 expts.)
Red = Effluent from 4ft. Coke bed (--- Average of 8 Expts.)
Blue = Effluent
X = No record for this date (4ft. Coke-bed effluent)

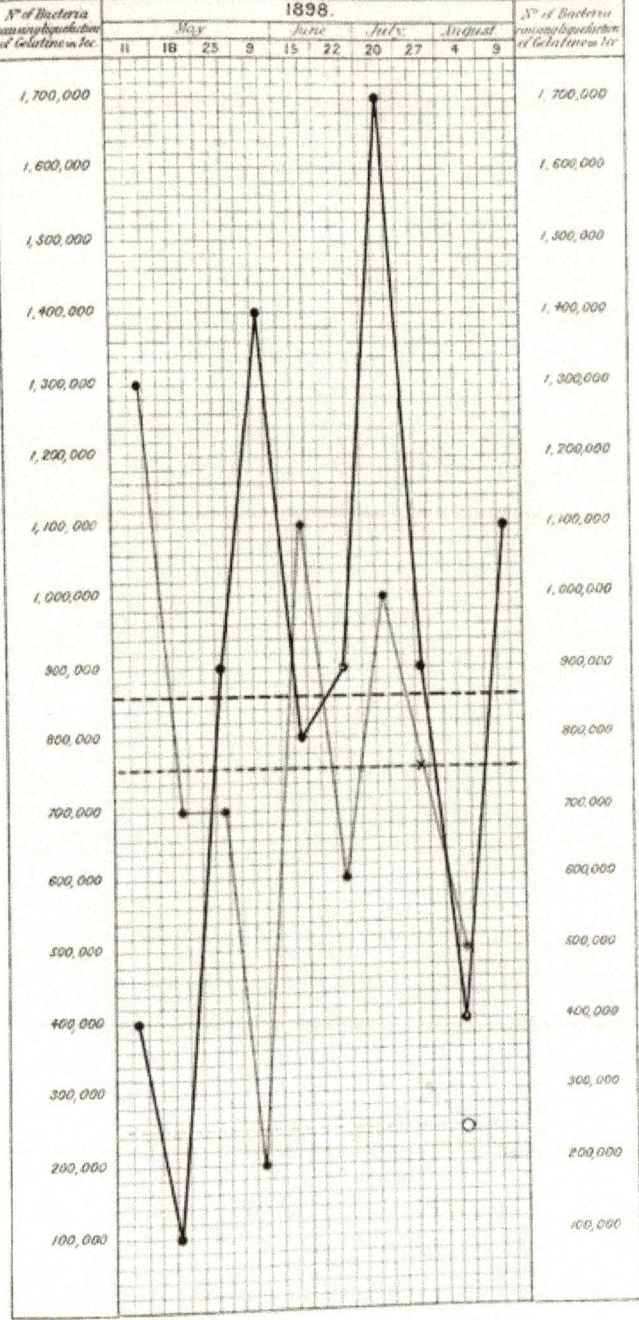

Diagram 5, showing the number of bacteria causing liquefaction of gelatine in 1 cc. of Crossness crude sewage (ten samples); in 1 cc. of effluent from 4-foot coke-bed (eight samples); and in 1 cc. of effluent from 6-foot coke-bed (two samples).

Diagram 6. showing, as regards Croasness crude sewage, the percentage deviation from the mean in 10 samples examined bacteriologically for the total number of bacteria, the number of spores of bacteria, and the number of liquefying bacteria.

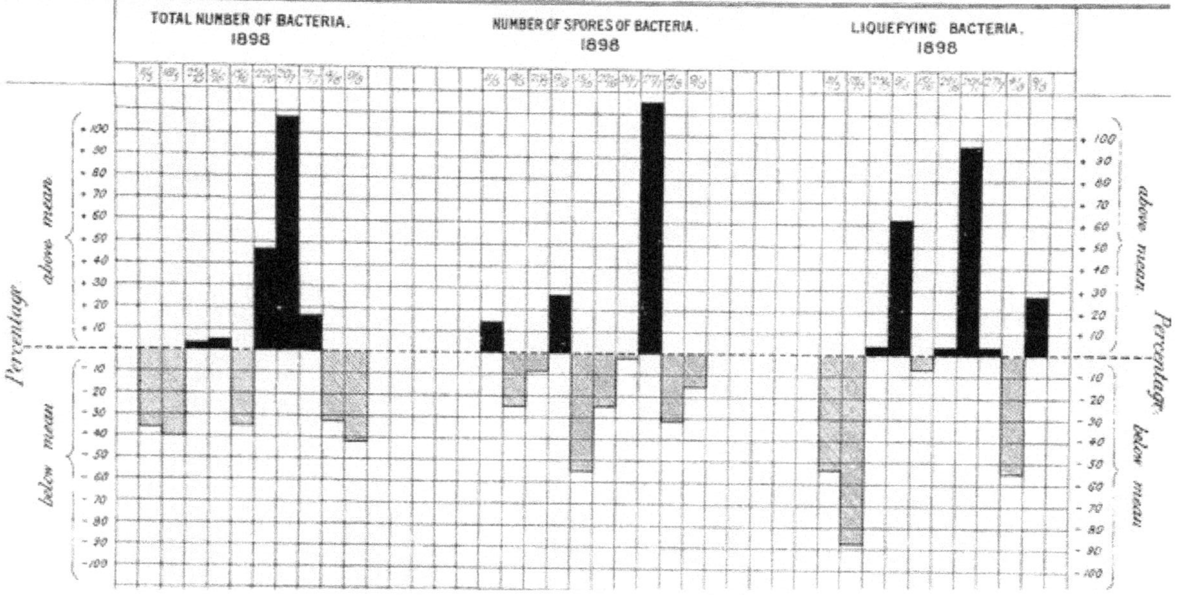

Diagram 7, showing, as regards the effluents from the 4-foot and 6-foot coke-beds, the percentage deviation from the mean in eight samples and two samples respectively, examined bacteriologically for the total number of bacteria, the number of spores of bacteria, and the number of liquefying bacteria.

▨ = 4 ft. Coke-bed effluent.   ☐ = 6 ft. Coke-bed effluent.

Diagram 8, showing the number of spores of B. Enteritidis Sporogenes (Klein) in 1 cc. of Crossness crude sewage (eleven samples); in 1 cc. of effluent from 4-foot coke-bed (ten samples); in 1 cc. of effluent from 6-foot coke-bed (five samples); and in 1 cc. of effluent from laboratory coke-bed (effluent from 6-foot coke-bed again treated in laboratory at Crossness—four samples).

Number of Spores of B. Enteritidis in 1 cc.	1898.										Number of Spores of B. Enteritidis in 1 cc.
	May			June			July			August		
	11	18	25	9	15	22	6	20	27	4	9	
1100												1100
1000												1000

Diagram 9, showing the number of B. Coli (or closely allied forms) in 1 cc. of Crossness crude sewage (ten samples); in 1 cc. of effluent from 4-foot coke-bed (eight samples); and in 1 cc. of effluent from 6-foot coke-bed (two samples). [Calculated from the number of colonies in phenol gelatine plate culture containing 0·00001 cc. sample, which were indistinguishable from B. Coli in the characters of their growth.]

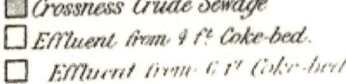

■ Crossness Crude Sewage
□ Effluent from 4 ft. Coke-bed.
□ Effluent from 6 ft. Coke-bed.

PLATE I.

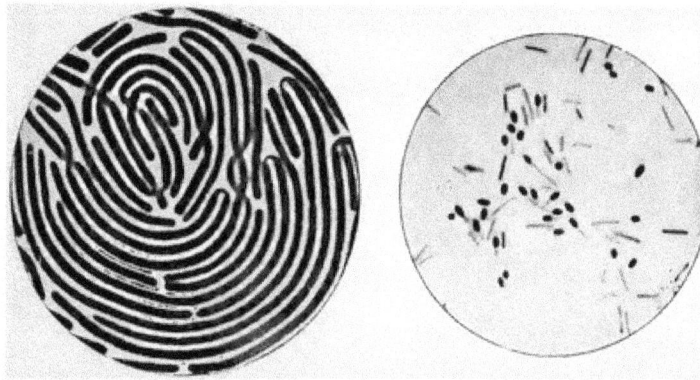

FIG. 1—Proteus vulgaris. Impression preparation from "swarming islands" on gelatine; 20 hours' growth at 20° C. × 3,000.

FIG. 2—B. enteritidis sporogenes (Klein). Microscopic double-stained preparation, from a serum culture, showing spores × 2,000.

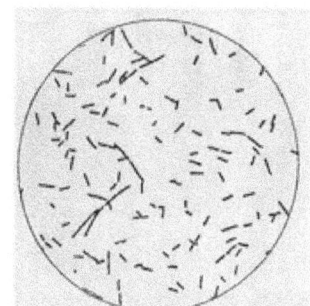
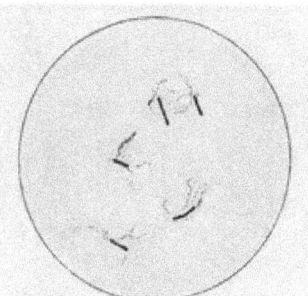

FIG. 3—B. mesentericus. Sewage variety E. Microscopic preparation from a 20 hours' agar culture at 20° C. × 1,000.

FIG. 4—B. mesentericus. Sewage variety E. Microscopic preparation stained by V. Ermengem's method, showing numerous flagella, from a 20 hours' agar culture at 20° C. × 1,000.

PLATE II.

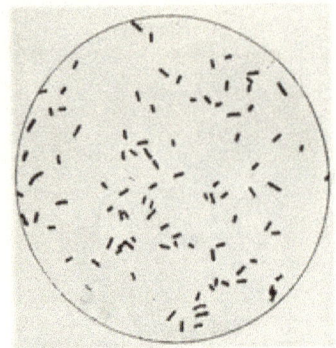

Fig. 5—B. mesentericus. Sewage variety I. Microscopic preparation from a 20 hours' agar culture at 20° C. × 1,000.

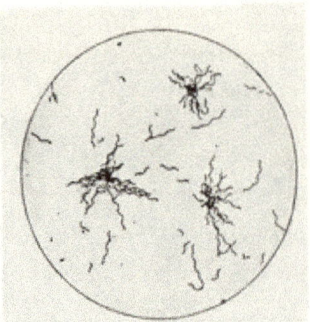

Fig. 6—B. mesentericus. Sewage variety I. Microscopic preparation stained by V. Ermengem's method, showing numerous flagella; from a 20 hours' agar culture at 20° C. × 1,000.

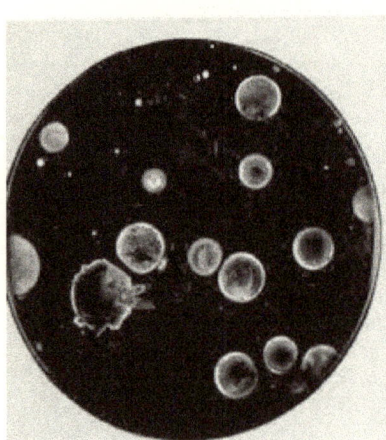

Fig. 7—B. mesentericus. Sewage variety I. Gelatine plate culture, about natural size.

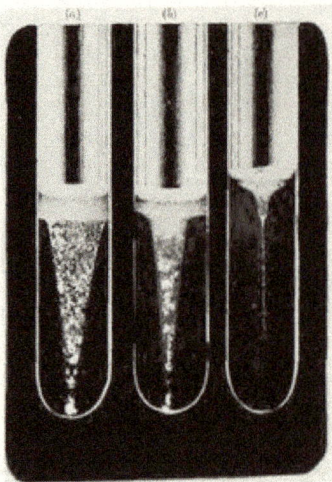

Fig. 8—B. mesentericus. Sewage variety I. Gelatine "stab" cultures, about natural size—
(a) Three days' growth at 20° C.
(b) Two days' growth at 30° C.
(c) One day's growth at 20° C.

PLATE III.

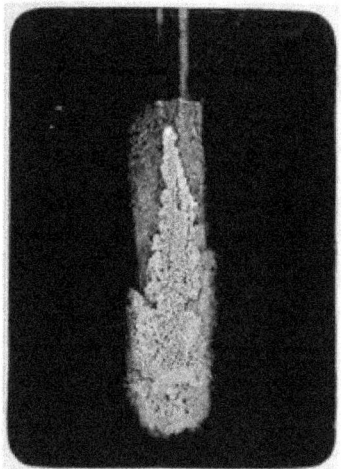

FIG. 9—B. mesentericus. Sewage variety I. Potato culture, one day's growth at 37° C., very slightly enlarged.

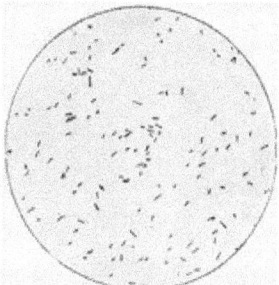

FIG. 10—"Sewage proteus." Microscopic preparation from an agar culture, 24 hours' growth at 20° C. × 1,000.

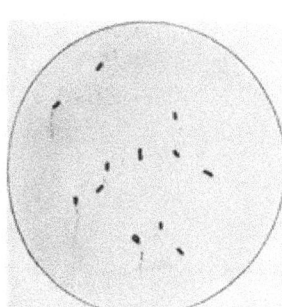

FIG. 11 — "Sewage proteus." Microscopic preparation stained by V. Ermengem's method, showing one flagellum at the end of each rod; from a 24 hours' growth agar culture at 20° C. × 1,000.

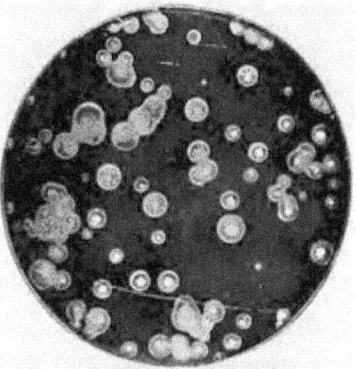

FIG. 12—"Sewage proteus." Gelatine plate culture, two days' growth at 20° C., about natural size.

PLATE IV.

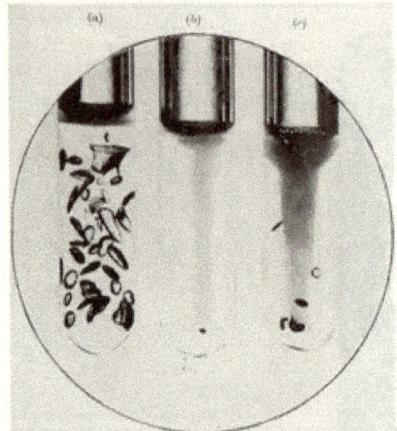
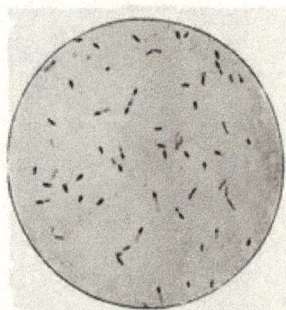

FIG. 14.—B. frondosus. Microscopic double-stained preparation from an agar culture showing spores × 1,000.

FIG. 13.—"Sewage proteus." About natural size—
(a) Gelatine "shake" culture. 24 hours' growth at 20° C.
(b) Gelatine "stab" culture. 24 hours' growth at 20° C.
(c) Gelatine "stab" culture. 48 hours' growth at 20° C.

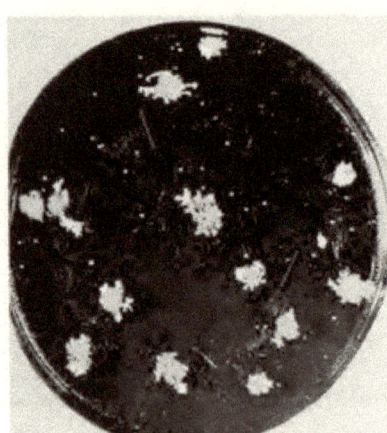
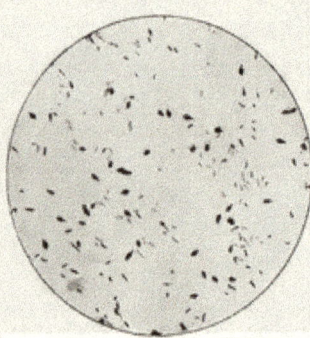

FIG. 16.—B. fusiformis. Microscopic double-stained preparation, showing spores × 1,000.

FIG. 15.—B. frondosus. Gelatine plate culture, about natural size.

PLATE V.

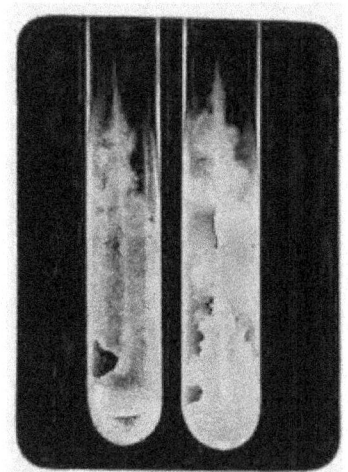

Fig. 17.—B. subtilissimus. Impression preparation from a gelatine plate culture × 1,000.

Fig. 18.—B. subtilissimus. Gelatine "streak" cultures, 24 hours' growth at 20° C., natural size.

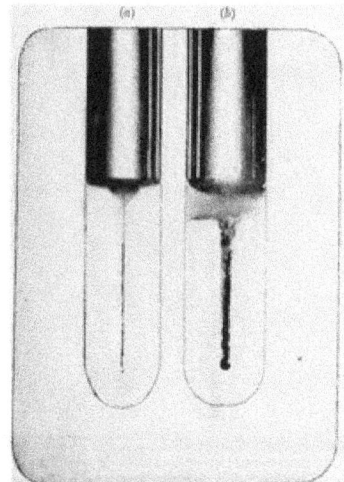

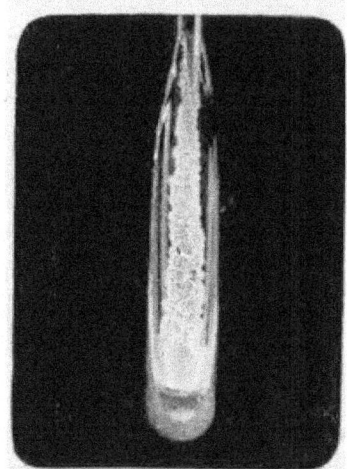

Fig. 19.—B. subtilis. Sewage variety B. Gelatine "stab" cultures, about natural size—
(a) One day's growth at 20° C.
(b) Three days' growth at 20° C.

Fig. 20.—B. subtilis. Sewage variety B. Oblique agar culture, 3 days' growth at 20° C., about natural size.

PLATE VI.

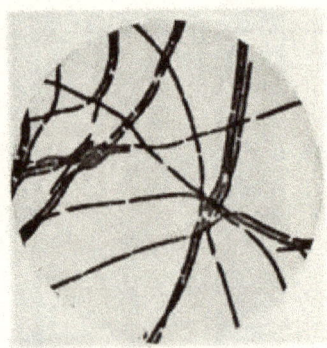

FIG. 21—B. membraneus patulus. Impression preparation from a gelatine plate culture × 1,000.

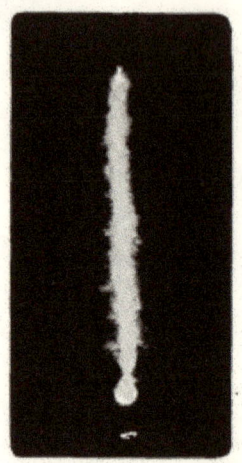

FIG. 22.—B. membraneus patulus. Oblique gelatine culture, about natural size.

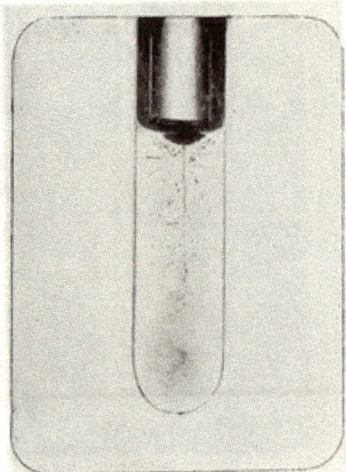

FIG. 23—B. membraneus patulus. Gelatine "stab" culture, 3 days' growth at 20° C., about natural size.

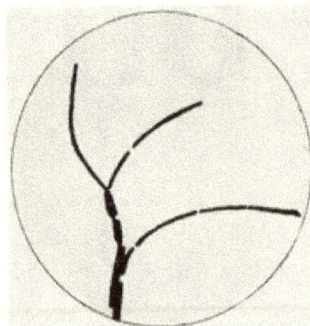

FIG. 24—B. capillareus. Impression preparation from a gelatine plate culture, 20 hours' growth at 20° C. × 1,000.

PLATE VII

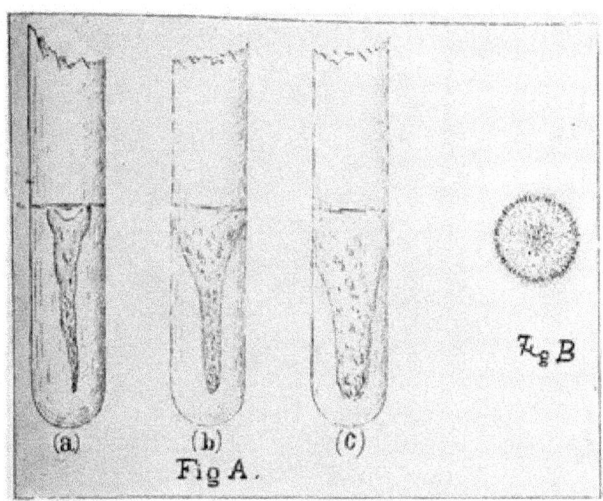

FIG. A.—B. subtilis. Sewage variety A. Gelatine "stab" cultures—
(a) Two days' growth at 20° C. (b) Three days' growth at 20° C. (c) Four days' growth at 20° C.
FIG. B.—B. subtilis. Sewage variety A. Colony in a gelatine plate under a low power of the microscope. Two days' growth at 20° C.

(Diagrammatic.)

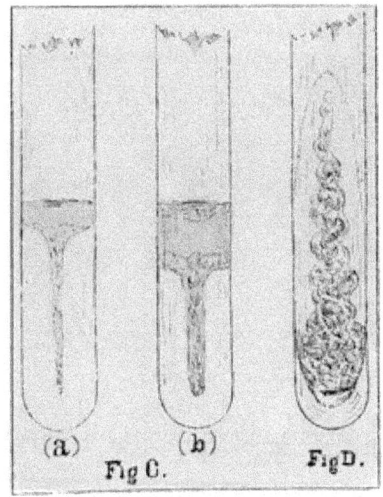

FIG. C.—B. subtilis. Sewage variety B. Gelatine "stab" cultures—
(a) Two days' growth at 20° C.
(b) The same at a later stage.
FIG. D.—B. subtilis. Sewage variety B. Blood serum culture. Forty-eight hours' growth at 37° C.

(Diagrammatic.)

PLATE VIII.

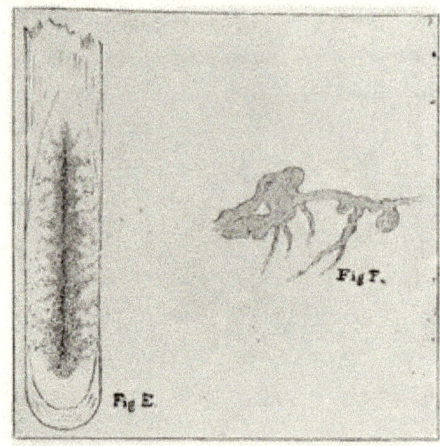

FIG. E.—B. membraneus patulus. Oblique gelatine culture.
FIG. F.—B. membraneus patulus. Showing the appearance under a low power of the microscope, of the delicate film-like processes which extend over the surface of the medium in gelatine plate cultures.

(Diagrammatic.)

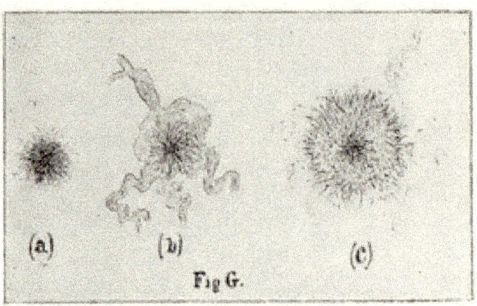

FIG. G.—B. capillareus. Under a low power of the microscope—
(a) Deep-seated colony in gelatine. Twenty hours' growth at 20° C.
(b) Colony partly deep and partly superficial. Twenty hours' growth at 20° C.
(c) Colony at a later stage of growth and after liquefaction had set in.

(Diagrammatic.)

www.ingramcontent.com/pod-product-compliance
Lightning Source LLC
Chambersburg PA
CBHW020243090426
42735CB00010B/1819